Science and

Mathematics for

This book is to be returned on or before

Engineering

£14·99

To the memory of my father, George William Stothard, who introduced me to engineering

Science and Mathematics for Engineering
Intermediate GNVQ Workbook

Dave Stothard
Macmillan College,
Middlesborough

A member of the Hodder Headline Group
LONDON • SYDNEY • AUCKLAND

First published in Great Britain in 1999 by
Arnold, a member of the Hodder Headline Group,
338 Euston Road, London, NW1 3BH

http://www.arnoldpublishers.com

British Library Cataloguing in Publication Data
A catalogue record for this book is available from the British Library

ISBN 0 340 70005 X

1 2 3 4 5 6 7 8 9 10

Commissioning Editor: Sian Jones
Production Editor: Rada Radojicic
Production Controller: Priya Gohil
Cover Design: Mouse Mat Design

Typeset in 10/12pt Stone Serif Regular and Frutiger Condensed
by J&L Composition Ltd, Filey, North Yorkshire
Printed and bound in Great Britain by J. W. Arrowsmith Ltd, Bristol

What do you think about this book? Or any other
Arnold title? Please send your comments to
feedback.arnold@hodder.co.uk

Contents

Preface

This book is designed to provide essential coverage of the elements and performance criteria for Unit 3: Science and Mathematics for Engineering, for GNVQ Intermediate Engineering.

Chapter 1 introduces the SI system of units and the basic physical quantities used in engineering, while chapter 2 reviews the mathematical techniques involved in making calculations. Each of chapters 3 to 9 investigates the scientific laws and principles for a particular type of engineering system and uses mathematical techniques to calculate relationships between physical quantities. An assessment activity at the end of each of chapters 3 to 9 gives students the opportunity to apply the principles in a systematic approach through experiments and investigations. Within these activities key skills in application of number and information technology can be achieved and referenced in portfolios of evidence. The book concludes with a set of multiple choice questions similar to those encountered in the GNVQ examination.

This book draws on the practical experience gained through teaching GNVQ Intermediate Engineering at the City of Sunderland College from the time the qualification was first introduced. It is hoped that both students and teachers will find it helpful. The author would like to thank Sunderland College for the experience and inspiration to write this book.

Dave Stothard
Billingham
June 1998

chapter

1 Introduction

1.1 The SI system of units

In 1960 a new system of units was introduced which has rapidly gained recognition worldwide. This International System of Units (Système international d'unités), abbreviated to SI, is based on the seven physical quantities and their units listed in Table 1.1.

Table 1.1 SI base units

Physical quantity	Name of unit	Symbol
Length	metre	m
Mass	kilogram	kg
Time	second	s
Electric current	ampere	A
Thermodynamic temperature	kelvin	K
Luminous intensity	candela	cd
Amount of substance	mole	mol

Prefixes

Engineering science and mathematics often involve very large or very small numbers. Prefixes involving powers of 10 which are a multiple of three are often used. Table 1.2 shows the common prefixes used.

Table 1.2 SI unit prefixes

Multiple or submultiple		Prefix	Symbol
10^{12}	= 1 000 000 000 000	tera	T
10^{9}	= 1 000 000 000	giga	G
10^{6}	= 1 000 000	mega	M
10^{3}	= 1 000	kilo	k
10^{2}	= 100	hecto	h
10^{1}	= 10	deca	da
	1 unit	–	–
10^{-1}	= 0.1	deci	d
10^{-2}	= 0.01	centi	c
10^{-3}	= 0.001	milli	m
10^{-6}	= 0.000 001	micro	μ

Exercise 1.1

1 Complete the following table, using the engineering systems in this book as a guide.

Quantity	Name of unit	Symbol
Length	metre	m
Mass		
Time		
Electric current		
Thermodynamic temperature		
Force		
Velocity		
Acceleration		
Potential difference		

2 Complete the table of multiples and submultiples for these common prefixes.

Multiplication factor	Power	Prefix	Symbol
1000	10^3	kilo	k
		mega	
		milli	
		giga	
		micro	
		centi	

3 Convert the following quantities:

(a) 600 mm to metres =
(b) 7200 g to kg =
(c) 240 m to km =
(d) 0.5 m to mm =
(e) 2500 kg to tonnes =
(f) 650 cm to m =
(g) 1.2 kg to g =
(h) 0.75 m to mm =

1.2 Physical quantities

The physical quantities that are most important in engineering are length, mass, area and volume.

Length is the distance between two points. The distance governs the unit to use. For example a small bolt would be measured in millimetres, possibly centimetres, but the metre unit would be inappropriate. A room would be measured in metres, rather than centimetres.

Mass is the amount of matter in a body. The mass of a large body would be measured in kilograms or tonnes, while the mass of a smaller body would be measured in grams.

1000 g = 1 kg
1000 kg = 1 tonne

Area is a measure of the size of a plane surface and for a rectangle is found by multiplying the length of one side by the length of the other. Large areas are normally measured in square metres, for example, a workshop measuring 10 m by 6 m has an area of $10 \times 6 = 60 \text{ m}^2$

The following formulae can be used to find the areas of regular shapes such as rectangles, squares, triangles and circles.

Area of a rectangle = $b \times h$

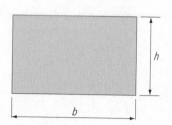

Area of a square = l^2 since the sides are the same length

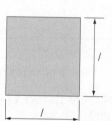

Area of a circle = $\dfrac{\pi D^2}{4}$ where D is the diameter,

or area = πr^2 where r is the radius

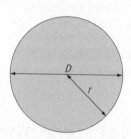

Area of a hollow circle = $\dfrac{\pi}{4}(D^2 - d^2)$,

where D is the large diameter and d is the small diameter

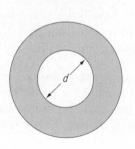

Area of a triangle =
½ base × height

= $\frac{1}{2}\,bh$

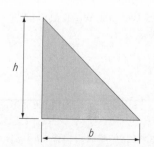

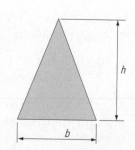

Volume is the measure of the space a body occupies. The volume of a body is found by multiplying the area by the height. Large volumes are measured in cubic metres, for example, if a packing crate 3 m × 2 m is 1.5 m high, it has a volume of 3 × 2 × 1.5 = 9 m^3. If the volume of a liquid is to be measured, the litre is used, where 1000 cm^3 = 1 litre.

The following formulae can be used to find the volumes of regular cuboids or cylinders.

Volume of a cube or block = b 3 h 3 d

where d is the depth of the block, b is the breadth, and h is the height

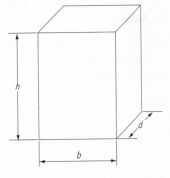

Volume of a cylinder = cross-sectional area × length (or height)

$$= \frac{\pi D^2}{4} \times h$$

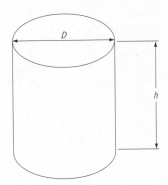

EXAMPLE 1.1

Calculate the volume of a cube of side 50 mm.

Volume = base × height × depth

$= 50 \times 50 \times 50$

$= 125\ 000$ mm^3

The volume of a cube of side 50 mm is 125 000 mm^3.

Note that the unit mm^3 can be used to measure a small volume such as the cube in this example.

EXAMPLE 1.2

Calculate the volume of a cylindrical air tank 2.2 m in diameter and 3.5 m high.

volume = area × height

$= \frac{\pi D^2}{4} \times 3.5$

$$= \frac{\pi \times 2.2^2}{4} \times 3.5$$

$$= 3.8 \times 3.5$$

$$= 13.3 \text{ m}^3$$

The volume of the tank is 13.3 m³.

EXAMPLE 1.3

Each cylinder in a four-cylinder car engine has a bore (diameter) of 73.7 mm and a stroke (length) of 87.5 mm (Fig. 1.1). Determine the volume of the cylinder in cm³ and the total capacity of the engine in litres.

Figure 1.1 Engine cylinder.

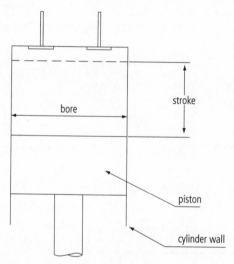

Volume of one cylinder = area × stroke

$$\frac{\pi D^2}{4} \times 8.75$$

$$= \frac{\pi \times 7.37^2}{4} \times 8.75$$

$$= 42.66 \times 8.75$$

$$= 373.28$$

The volume of one cylinder is 373.28 cm³

The total capacity of the engine = volume of one cylinder × number of cylinders

$$= 373.28 \times 4$$

$$= 1493 \text{ cm}^3$$

If this value is expressed in litres, where 1 litre = 1000 cm³, then the capacity of the engine can be found by dividing 1493 by 1000, i.e.

$$\frac{1493}{1000} = 1.493$$

$$= 1.5 \text{ litres}$$

The total capacity of the engine is 1.5 litres.

EXAMPLE 1.4

Calculate the volume of the mild steel plate shown in Fig. 1.2 which was produced from an original piece of dimensions 80 mm long, 50 mm wide and 3 mm thick.

Figure 1.2 Mild steel plate.

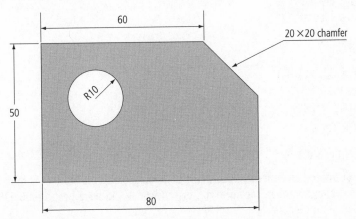

The first step is to find the area of the plate. Calculating the area of the original 80 × 50 plate and subtracting the triangular area and the hole will leave the area of the completed plate .

Original area of metal = 80 × 50 = 4000 mm^2

Area of triangular chamfer = ½ base × height

$$= \frac{1}{2} \times 20 \times 20$$

$$= 10 \times 20$$

$$= 200 \text{ mm}^2$$

Area of hole = $\frac{\pi D^2}{4}$

$$= \frac{\pi \times 20^2}{4}$$

$$= 314.2 \text{ mm}^2$$

Area of plate = original area − area of chamfer − area of triangle

$$= 4000 - 200 - 314.2$$

$$= 3485.8$$

The area of the plate is 3485.8 mm^2.

The volume of the plate can now be calculated as follows.

Volume of plate = area × thickness

$$= 3485.8 \times 3$$

$$= 10\ 457.4\ mm^3$$

The volume of the plate is 10 457.4 mm³.

EXAMPLE 1.5

Calculate the volume of material in a mild steel rod 20 mm diameter and 2 m long.

(Note: the diameter and the length must be in the same units. As the diameter is small in comparison to the length, the units have been shown in mm.)

Volume = area x length

$$= \frac{\pi D^2}{4} \times 2000$$

$$= \frac{\pi \times 20^2}{4} \times 2000$$

$$= 314.2 \times 2000$$

$$= 628\ 318.5$$

The volume of the rod is 628 318.5 mm³.

The volume of the rod looks large, but it must be remembered that the unit is mm³ (cubic millimetres). If the value had been shown in m³, the answer would have been 0.000 628 3m³, since there are 10^9 mm³ in every m³.

Exercise 1.2

1 A fitting exercise for an apprentice requires the plate shown in Fig. 1.3 to be produced from flat bar of dimensions 75 mm × 50 mm × 3 mm thick. For the completed plate determine:

(a) the volume of material in the plate
(b) the volume of material removed as a percentage of the original flat bar.

Draw a pie chart to show the amount of material removed from the original bar.

Figure 1.3 Plate made from 3 mm thick flat bar.

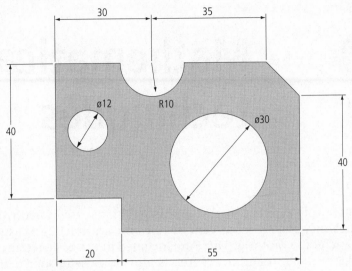

Note: Ø represents the diameter

2 A workshop stores has quantities of mild steel bar in the sizes and lengths shown. Determine the total volume of mild steel and show which of the quantities has the largest volume.

Diameter (mm)	Length (m)	Quantity
10	2.3	20
12	3.1	12
16	0.6	10
20	3	10
30	3	6

3 A sports car engine has the following cylinder dimensions: 88.9 mm bore, 71.1 mm stroke. If the engine has eight cylinders, determine the engine capacity in both cubic centimetres and litres.

chapter

2 Mathematical techniques

This chapter deals with the mathematical techniques associated with science and engineering. It covers the general rules for arithmetic, transposition of formulae and graph plotting. These are included in the basic requirements for element 3.3 of the unit and the methods provide a useful source of information for problem solving throughout the unit.

2.1 BODMAS

The basic arithmetic operations are addition $(+)$, subtraction $(-)$, multiplication $(\times)$ and division $(\div)$. Often a problem involves several of these operations. The BODMAS rule can be used to decide which operation to do first.

BODMAS stands for Brackets, Of, Divide, Multiply, Add, Subtract, and the arithmetic operations in a problem should be approached in that order. It should be noted that 'of' can also mean 'multiply'. The following examples show how the BODMAS rule is used.

EXAMPLE 2.1

$$6 + 4 \times (5 - 3) = 6 + 4 \times 2 \qquad \text{Brackets}$$
$$= 6 + 8 \qquad \text{Multiply}$$
$$= 14 \qquad \text{Add}$$
$$\text{answer} = 14$$

EXAMPLE 2.2

$$3 + 5 \times 6 = 3 + 30 \qquad \text{Multiply}$$
$$= 33 \qquad \text{Add}$$
$$\text{answer} = 33$$

EXAMPLE 2.3

$6 \div 3 \times (12 - 4) = 6 \div 3 \times 8$ Brackets

$= 6 \div 24$ Multiply

$= 0.25$ Divide

answer $= 0.25$

Exercise 2.1

Try the following problems using the BODMAS rule as in the examples.

(a) $26 - (3 + 2)$

(b) $22 - (8 + 7)$

(c) $16 - (8 + 3)$

(d) $5 + (15 \div 5)$

(e) $6 + (9 \div 3)$

(f) $4 \times (2 + 3)$

(g) $3 \times (3 + 2)$

(h) $15 \times (3 + 2)$

The times sign is not required when used in front of brackets, as in (f), (g) and (h) of Exercise 2.1. For example,

$6 \times (4 + 3)$ is the same as $6(4 + 3)$

In this type of question we will work out the bracket first, then multiply by 6:

$6(4 + 3) = 6 \times 7$

$= 42$

Exercise 2.2

Solve the following

(a) $4(3 + 2)$

(b) $12(9 - 6)$

(c) $3(6 + 9)$

(d) $6(9 - 4)$

(e) $6(5 - 2)$

(f) $5(12 - 3)$

(g) $3(16 \div 4)$

(h) $4(21 \div 3)$

Check your answers to Exercises 2.1 and 2.2 with the answers given at the end of the book

In some problems there may not be any brackets, but the BODMAS rule can still be used. For example,

$12 + 3 \times 6$

Remember, BODMAS tells us to do the division and multiplication before the addition.

- $12 + 3 \times 6$

  $12 + 18 = 30$

- $5 + 4 \times 2$

  $5 + 8 = 13$

Exercise 2.3

Try the following:

(a) $5 + 3 \times 6$

(b) $7 + 2 \times 5$

(c) $15 + 8 \times 3$

(d) $4 + 6 \times 7$

(e) $3 + 4 \times 3$

(f) $6 + 12 \div 3$

(g) $7 + 24 \div 6$

(h) $12 + 27 \div 3$

Check your answers.

Exercise 2.4

Using the BODMAS rule, work out the following and check your answers.

(a) $3 + 4(5 - 2)$

(b) $2 \times 4 + 6 \times 3$

(c) $5 \times 3(6 - 2)$

(d) $19 - 4(3 \times 2)$

(e) $(12 - 2) \div (6 - 4)$

(f) $(4 \times 3) \times (5 - 2)$

(g) $16 - 4(5 - 2)$

(h) $12 \div 3 \times 4$

Section review

- BODMAS stands for Brackets, Of, Divide, Multiply, Add, Subtract
- Brackets must be worked out first
- If there are no brackets in the problem, go to the next stage in the rule.

2.2 Arithmetic

This section deals with the methods of addition, subtraction, division and multiplication of numbers. These standard operations are normally performed using a calculator, but there is still the need to solve problems accurately without using a calculator. It is worthwhile remembering that aptitude tests often have an arithmetic section which must be completed without using a calculator.

In arithmetic the following symbols are used:

+ for add $\quad\quad$ 4 + 5

− for subtract $\quad$ 12 − 5

÷ for divide $\quad$ 24 ÷ 3

× for multiply $\quad$ 6 × 5

Addition

4 + 3 = 7

It is rare that an addition of two numbers is as straightforward as this. However, the most complex of numbers, large or small, can be broken down so that each step can be quite simple.

EXAMPLE 2.4

Find the sum of 26 + 7 + 124 + 86

Method

The numbers are placed in columns of units, tens, hundreds and thousands (if any). Units are single numbers, i.e. less than 10.

$$
\begin{array}{r}
26 \\
7 \\
124 \\
\underline{86} \\
243 \\
\hline
{\scriptstyle 1\ 2}
\end{array}
$$

Start by adding the units column. Thus 6 and 7 make 13, plus 4 makes 17, and plus 6 makes 23. Write the 3 as the total in the units column, and carry the 2 forward to the tens column. Now add the tens. Thus 2 plus 2 makes 4, plus 8 makes 12, plus the 2 carried forward makes 14. Put the 4 in the tens column of the total and carry the 1 forward to the hundreds column. Now add the hundreds column. The 1 already in the hundreds column plus the 1 carried forward makes 2. Place this number in the hundreds column to show the answer of 243.

EXAMPLE 2.5

Find the value of 45 + 1245 + 2456 + 27

$$
\begin{array}{r}
45 \\
1245 \\
2456 \\
27 \\
\hline
3773 \\
\hline
{\scriptstyle 1\ 2}
\end{array}
$$

- Add the units: 5 plus 5 plus 6 plus 7 makes 23. Place the 3 in the units column of the total and carry the 2 forward to the tens column.
- Add the tens: 4 plus 4 (is 8) plus 5 (is 13) plus 2 (is 15), plus the 2 carried forward makes 17. Place the 7 in the tens column of the total and carry the 1 forward to the hundreds column.
- Add the hundreds: 2 plus 4 (is 6) plus the 1 carried forward makes 7. Place the 7 in the hundreds column of the total.
- Add the thousands: 1 plus 2 makes 3. Place the 3 in the thousands column to show the total of 3773.

EXAMPLE 2.6

A mechanical component is shown in Fig. 2.1. The dimensions are in mm. Determine the overall length L shown.

Figure 2.1   Mechanical component.

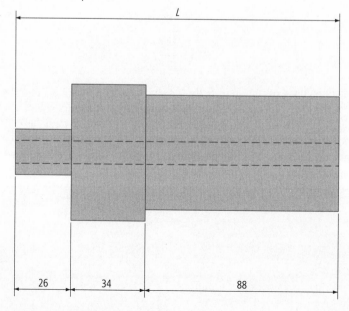

length $L = 26 + 34 + 88 = 148$ mm

Exercise 2.5

1 Find the value of the following:

(a) 25 + 13 + 56

(b) 45 + 17 + 125

(c) 145 + 16 + 1029

(d) 1245 + 212 + 42 + 65

2 Find the total length in mm of the shaft shown in Fig. 2.2.

Figure 2.2 Shaft.

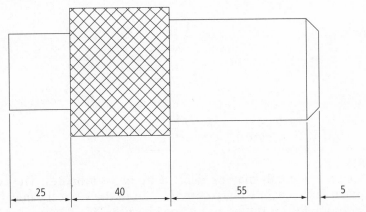

3 Four masses of 26 kg, 45 kg, 16 kg and 73 kg are placed on a bench. Find the value of the total mass.

4 A store contains 20 mm diameter mild steel bar of the following lengths: 800 cm, 250 cm, 344 cm, 234 cm and 98 cm. Find the total length of the steel bar in cm.

5 A forklift truck carries a crate containing six boxes. Each box has ten components, each component mass being 20 kg. Determine the total mass carried.

6 A lift is occupied by four people with the following weights: 70 kg, 120 kg, 90 kg and 140 kg. Determine the mass the lift is carrying. If the maximum load the lift can carry is 500 kg, determine the maximum weight of a fifth person.

2.3 Graphs – general principles

There are many types of graph. Most use two coordinates (usually x and y) plotted on two axes at right angles. The independent variable is normally x, which is plotted horizontally, and the dependent variable, y, which is plotted vertically. The point at which the axes intersect is called the origin. Distances to the right and above the origin are positive values, while distances to the left or below the origin are negative values (Fig. 2.3).

Figure 2.3   Features of a graph.

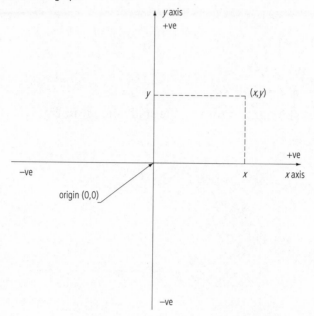

Any point on the graph may be defined by its coordinates. The horizontal value is given first, separated by a comma ',' from the vertical value. For example, the coordinates for the point on the graph in Fig. 2.4. are (7, 5) where x = 7 and y = 5.

Figure 2.4   Plotting a point on a graph.

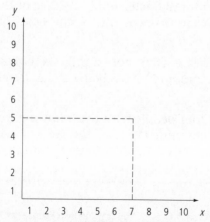

To decide which variable is the dependent or independent, try to think of **which depends on which.** For example, the temperature of a cup of tea depends on the time since it was poured out. Temperature is the dependent variable here and time the independent variable.

The graph showing how the temperature of a cup of tea depends on time after pouring would be plotted with time on the horizontal axis and temperature on the vertical axis, as shown below.

Figure 2.5 Change in temperature of a cup of tea with time.

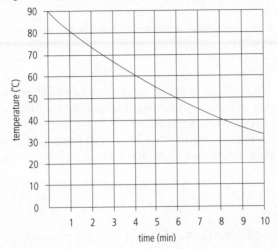

From this graph, the temperature is 50°C after cooling for 6 minutes, so the coordinates of this point are (6, 50).

Table of values

A table of values should be drawn up before plotting the graph. The values may be the results from an experiment, or in the case of statistical data, they may require grouping to produce a table.

An example of a set of results for an object moving on a conveyor belt is shown below.

Distance (m)	0	10	15	20	25	30
Time (s)	0	5	10	15	20	25

If an equation is to be plotted, then a systematic approach is required to produce the values to be plotted.

For example: plot the graph of $y = 2x + 3$ for the range -2 to $+6$.

The x values are required to be in the range -2 to $+6$. The y values are found in the following way.

x	-2	-1	0	1	2	3	4	5	6
$2x$	-4	-2	0	2	4	6	8	10	12
$+3$	$+3$	$+3$	$+3$	$+3$	$+3$	$+3$	$+3$	$+3$	$+3$
y	-1	1	3	5	7	9	11	13	15

In the table above:

- the top row shows the x value
- the second row shows $2 \times x$
- the third row adds 3
- the bottom row gives the total for $y = 2x + 3$ for each x value

Scales

Scales should be chosen which make as much use of the paper space as possible. Each large graph square should represent 1, 2 or 5 units, or multiples of these numbers. The axes should be labelled with the name of the variable and the unit should be shown. A graph should always have a title, which usually describes the variables, such as a velocity–time graph or a force–extension graph.

Plotting the graph

The points on the graph should be plotted from the results table and marked with a dot or a small cross. A line is drawn through the points, or when the points are not in line, a best fit line is drawn.

The gradient of a graph

When a straight line graph passes though the origin then a direct proportional relationship exists between the variables (Fig. 2.6). For this type of graph the ratio of y/x will be a constant and is known as the slope or gradient. From this any y value can be found by multiplying the gradient and the x value, i.e. $y = mx$.

Figure 2.6 The gradient of a graph.

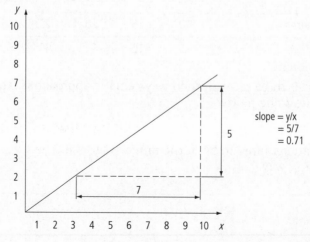

slope = y/x
= 5/7
= 0.71

General form of the straight-line graph

The equation of any straight-line graph can be expressed in the form $y = mx + c$, where m is the gradient of the line and c is the intercept on the y axis, the value of y when $x = 0$.

Figure 2.7 A straight-line graph.

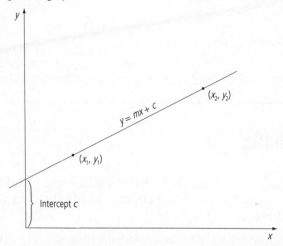

Exercise 2.6

1 The velocity of a car was measured at intervals of 5 seconds. Use the results given in the table to plot a graph of velocity against time for the car.

Velocity (m/s)	12	25	36	47	61	72
Time (s)	5	10	15	20	25	30

The gradient of this graph represents the acceleration of the car. Determine the gradient.

2 A load–extension test on a spring produced the following results. Plot the graph of load against extension and determine the gradient of the graph.

Load (N)	Extension (mm)
0	0
50	2
100	5
150	7
200	8.5
250	10.25

The gradient of this graph represents the amount of extension for each newton of applied force. This is termed the spring stiffness.

3 The temperature of a cooling liquid was measured at intervals of 5 minutes. The results are shown in the table. The initial temperature of the liquid was 67°C. Plot the temperature–time graph and determine the temperature of the liquid after 22.5 minutes.

Temperature (°C)	61	55	51	47.5	44	41.5
Time (min)	5	10	15	20	25	30

2.4   Pie charts

A pie chart is a graphical method of presenting information. The data that are used can be on any subject and the chart presents the data in a way that is clear and easy to read.

For example, the air we breathe is a mixture of gases, mainly nitrogen and oxygen, with small amounts of many others. The approximate amounts of the gases are: nitrogen, 4/5 or 78%; oxygen, 1/5 or 21%; noble gases, and carbon dioxide, remainder.

The composition can be presented visually using a pie chart, as shown in Fig. 2.8.

Figure 2.8   The composition of air.

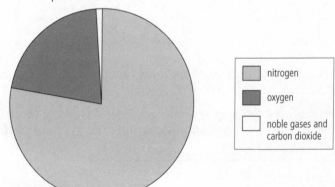

nitrogen

oxygen

noble gases and carbon dioxide

EXAMPLE 2.7

An engineering plant employs 60 staff. The numbers of staff employed in the various different job roles are: mechanical fitters, 5; electricians, 3; technicians, 2; stores, 2; production line, 34; administration, 6; and management, 8. To construct a pie chart to show this information, the percentage of workers in each job role is required.

Total number of employees = 60

Percentage of mechanical fitters = (5/60) × 100% = 8.33%

Percentage of electricians = (3/60) × 100% = 5%

Percentage of technicians = (2/60) × 100% = 3.33%

Percentage of stores = (2/60) × 100% = 3.33%

Percentage of production line = (34/60) × 100% = 56.67%

Percentage of administration = (6/60) × 100% = 10%

Percentage of management = (8/60) × 100% = 13.33%

The individual percentages can then be used to calculate how much of the pie chart each category will occupy. For example, since there are 360° in the pie, administration will occupy 10% of 360°, i.e. 36°.

An alternative method, which is very useful for information technology key skills, would be to input the original values into a spreadsheet and let the computer produce a pie chart like that in Fig. 2.9 from the many options in graphical packages.

Figure 2.9   Job roles of company employees.

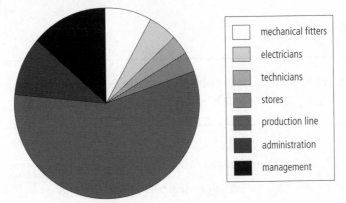

Spreadsheet packages are a very useful aid for presenting experimental data. Line, bar or column graphs and pie charts can all be produced and imported into word-processed reports to enhance the quality of presentation and demonstrate key skill competence.

2.5   Transposition of formulae

Mathematical formulae are often used to express the laws and principles of engineering science. The relationship between the different physical quantities is shown in the formula. For example, the relationship between voltage, current and resistance in an electric circuit is given by the formula

$$V = I \times R$$

where V is the voltage, I is the current and R is the resistance. If the voltage is required the formula can be used at once, as V is the subject of the formula. If the resistance or the current are required, the formula will need rearranging to make the desired quantity the subject. This is known as **transposition**.

The rules of transposition

RULE 1　If a term is multiplied or divided in the equation, it is transposed by dividing or multiplying both sides of the equation by the same quantity.

EXAMPLE 2.8

Transpose $V = IR$ to make R the subject.

$\dfrac{V}{I} = \dfrac{IR}{I}$　　divide each side of the equation by I to cancel I on the right hand side

$\dfrac{V}{I} = \dfrac{\cancel{I}R}{\cancel{I}}$　　the I on the bottom will cancel out the I on the top

$\therefore\quad \dfrac{V}{I} = R$

i.e.　$R = \dfrac{V}{I}$

RULE 2　If a term is added or subtracted in the equation, it is transposed by subtracting or adding the term to both sides of the equation.

EXAMPLE 2.9

Transpose $v = u + at$ to make u the subject.

$v - at = u + at - at$　　subtract at from both sides of the equation

$v - at = u + \cancel{at} - \cancel{at}$　　cancel out $at - at$

$\therefore\ v - at = u$　　or　　$u = v - at$

EXAMPLE 2.10

Transpose the formula $F = ma$ to make a the subject.

$\dfrac{F}{m} = \dfrac{\cancel{m}a}{\cancel{m}}$　　divide both sides by m and cancel out m on top and bottom lines

$\dfrac{F}{m} = a$

answer　$a = \dfrac{F}{m}$

EXAMPLE 2.11

Transpose the formula $y = mx + c$ to make x the subject.

$$y - c = mx + c - c \qquad \text{subtract } c \text{ from both sides}$$

$$\frac{y - c}{m} = \frac{mx}{m} \qquad \text{divide both sides by } m \text{ and cancel out}$$

$$\frac{y - c}{m} = x$$

answer $x = \dfrac{y - c}{m}$

Exercise 2.7

1 Transpose the formula $F = P \times A$ to make P the subject.

2 Transpose $v = a \times l$ to make l the subject.

3 Transpose $v = \dfrac{s}{t}$ to make s the subject.

4 Transpose $a = \dfrac{v - u}{t}$ to make u the subject.

5 Transpose $Q = mc\,(t_2 - t_1)$ to make t_1 the subject.

6 Transpose $y = mx + c$ to make m the subject.

7 Transpose $F = ma$ to make a the subject.

8 Rearrange $v = u + at$ to make t the subject.

9 If acceleration $= \dfrac{\text{velocity}}{\text{time}}$, rearrange to make time the subject.

10 Transpose $v^2 = u^2 + 2as$ to make s the subject.

chapter

3

Static systems

A system which remains at rest when a force is applied is known as a static system. The term 'at rest' means that the system does not move while under the effect of the applied force. In order to understand static systems, we must first define the term force.

3.1 Force

A force is defined as that which changes the state of rest of the body on which it acts. In a static system the force is counterbalanced and thus the body does not move. Should the counterbalance effect be removed, the body would no longer be at rest and would move.

In order to describe a force fully the following characteristics must be known:

- the **magnitude** which means the size
- the **direction** the line of action, whether horizontal, vertical or at a stated angle
- the **sense** the way the force acts on the line of action, whether up, down, left or right
- the **point of application** of the force

3.2 Scalar and vector quantities

A scalar quantity is one in which the magnitude (or size) only is given in the appropriate unit. Examples of scalar quantities are:

- a temperature of 30°C
- a mass of 20 kg
- a length of 20 m

A vector quantity has both magnitude and direction, for example:

- a force of 20 kN acting to the left of an object

This can be illustrated as in Fig. 3.1.

Figure 3.1 A force of 20 kN acting to the left of an object.

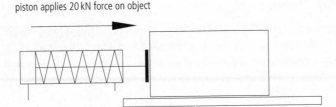

piston applies 20 kN force on object

3.3 Vectors

A vector is a line that completely represents a quantity. The force acting on the block in Fig. 3.1 can be represented as a vector.

Figure 3.1a

For the vector shown above:

- the length represents the magnitude
- the angle represents the direction (horizontal in this example)
- the arrow represents the sense (acting from left to right)
- the point of application is shown by the dot

Resultant vectors

If two vectors act at the same point in a body, then a resultant vector can be found which will produce the same effect as the two. The method used to find the value is known as the **parallelogram of forces**.

EXAMPLE 3.1

Forces A and B act at a point as shown in Fig. 3.2. Determine the single force which will produce the same effect as A and B using the parallelogram of forces.

Fig. 3.2 Parallelogram of forces.

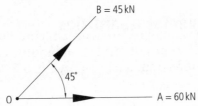

Method

1 Select a suitable scale, 1 mm = 1 kN

2 Draw A and B as vectors to scale

3 Draw a line parallel to OA at point B

4 Draw a line parallel to OB at point A to complete the parallelogram

5 Draw a diagonal line from O to R.

The vector diagram is shown in Fig. 3.3. Line OR represents the resultant force, the angle shows the direction and the arrow shows the sense. Thus a force of 97.18 kN acting at an angle of 19° to the horizontal will produce the same effect as forces A and B.

Figure 3.3 Vector diagram.

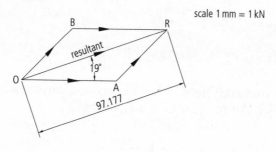

scale 1 mm = 1 kN

EXAMPLE 3.2

Two forces act at a point as shown in Fig. 3.4. Determine the resultant force using the parallelogram of forces method.

Figure 3.4 Resultant of two forces: (a) space diagram, (b) vector diagram.

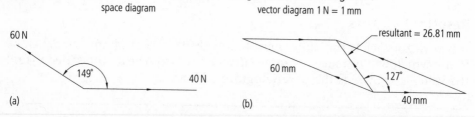

The method is shown in Fig. 3.4(b). The resultant force has a magnitude of 26.81 N and acts at an angle of 127° to the horizontal.

Equilibrant

The equilibrant is the force required to maintain equilibrium in a system, i.e. to keep the system static, or at rest. The equilibrant is equal to the resultant but acts in the opposite direction, as shown in Fig. 3.5.

Figure 3.5 Equilibrant of a force.

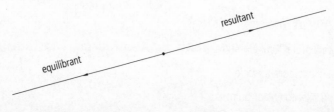

In the Example 3.2 the equilibrant would have been of magnitude 26.81 N but in the opposite direction, as shown in Fig. 3.6.

Figure 3.6 Equilibrant of the forces in Example 3.2.

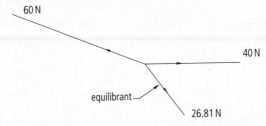

EXAMPLE 3.3

Two forces act at a point as shown in Fig. 3.7(a). Determine the magnitude and direction of the force required to maintain equilibrium in the system.

Figure 3.7 Finding the equilibrant: (a) space diagram, (b) vector diagram.

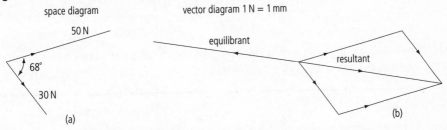

Figure 3.7(b) shows the vector diagram used to solve this question. For this system the required force can also be shown as in Fig. 3.8.

Figure 3.8 The system in equilibrium.

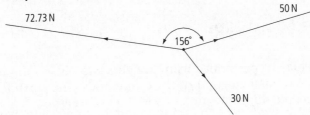

The three forces act on the same point and the system remains static. An alternative method for solving this type of problem is known as the **triangle of forces**.

The triangle of forces

In this method a vector is drawn representing a force. A second vector representing the second force is drawn at the end of the first vector, which produces two sides of a triangle. A third vector is then drawn to close the triangle. This vector represents the equilibrant.

EXAMPLE 3.4

Forces A and B act at a point as shown in Fig. 3.9(a). Determine the force required to maintain equilibrium in the system.

The method is shown in Fig. 3.9(b). From the triangle, the equilibrant force is 69.95 N.

Figure 3.9 Finding the equilibrant: (a) forces acting, (b) triangle of forces.

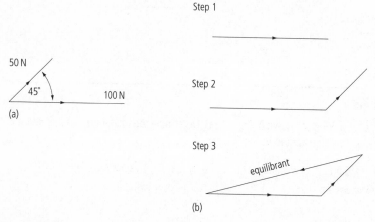

It should be noted that in a system of forces, where the vectors are drawn in this manner, a closed triangle indicates that the system will remain at rest.

EXAMPLE 3.5

Determine the equilibrant force for the situation shown in Fig. 3.10(a).

From the triangle of forces shown in Fig 3.10(b), the equilibrant force is 116 N acting 165° clockwise from the 40 N force, as shown in Fig. 3.10(c).

3.4 Force board experiments

An experiment may be performed using weights and pulleys to prove the triangle of forces and parallelogram of forces methods. The apparatus required is a wall-mounted board, some pulleys and a set of weights.

Experiment to find the equilibrant

The aim of the experiment is to determine the position and magnitude of the single force needed to maintain equilibrium in a system, using the parallelogram of forces.

Two forces of 80 N and 40 N act at right angles on a point as shown in Fig. 3.11.

Using the force board, determine the magnitude and direction of the force required to maintain equilibrium.

Figure 3.10 Finding the equilibrant: (a) forces acting, (b) triangle of forces, (c) space diagram.

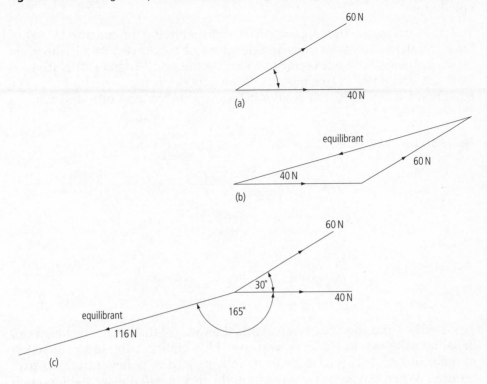

Figure 3.11 Forces acting at a point.

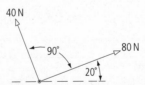

The two forces can be represented by weights attached to cords wound over pulleys positioned in the direction of the force. The two cords are joined to a ring, which is pegged into position on the board. The apparatus is set up as shown in Fig. 3.12.

Figure 3.12 Setting up the force board.

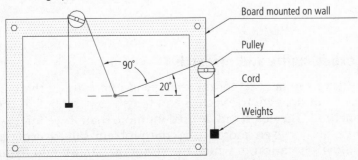

If the peg is removed at this stage the ring will move as the system is not in equilibrium.

The positions of the cords are recorded before removing the cords and weights. Lines are then drawn to show the direction of the forces. Parallel lines are drawn at the end of each vector to complete the parallelogram. The diagonal from the start point to the opposite corner represents the resultant force. The equilibrant is equal to the resultant but acts in the opposite direction, as shown in Fig. 3.13.

Figure 3.13 Finding the equilibrant using the parallelogram of forces.

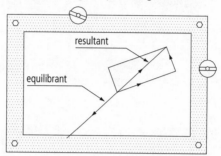

The length of the equilibrant represents the value of the force. The board can be set up as shown in Fig. 3.14, with suitable weights acting in the direction of the equilibrant. This is achieved by moving a third pulley into the correct position. When the peg is removed from the ring, it will remain static as equilibrium has been achieved.

Figure 3.14 Setting up the equilibrant on the force board.

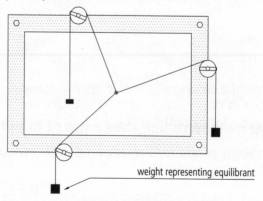

Further experiments and research

The force board can be used for the triangle of forces, using the same principle as above, but drawing the vectors end to end as described in the triangle of forces method. The board can be used for more than three forces – as many as four, five, six, or even more – and the problem solved using a method known as **Bow's notation** in which the vectors are drawn end to end.

Bow's notation

This is a method in which vectors are drawn end to end and forces are identified by letters placed within the space diagram (Fig. 3.15(a)). A closed vector diagram indicates that the system of forces is in equilibrium (Fig 3.15(b)).

Figure 3.15 Bow's notation.

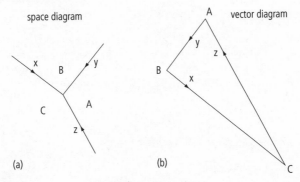

3.5 Centre of gravity

The centre of gravity of an object is the point through which its weight acts. For example, a plank of wood or rule will balance on a support placed at its centre of gravity, but if the support is moved to either side of the centre of gravity, then equilibrium cannot be maintained and the plank will tip (Fig. 3.16).

Figure 3.16 Balancing a plank.

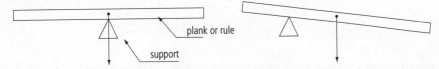

Experimental method to find the centre of gravity of a lamina

The centre of gravity of an irregular shaped thin sheet of plate, called a lamina, can be found by experiment. Two or three holes are drilled at regular intervals near the edge of the lamina, which is then suspended at a hole using a small rod or nail clamped to a stand (Fig. 3.17).

The lamina will rest with its centre of gravity vertically below the support point. A plumb line is hung from the support point to show the vertical, and a pencil line is drawn on the lamina to mark where the vertical lies. The lamina should now be rotated to the next hole and the process repeated. The point at which the lines intersect represents the centre of gravity. This can be confirmed by balancing the lamina on a small rod or finger, placed at this point.

Figure 3.17 Finding the centre of gravity of a lamina.

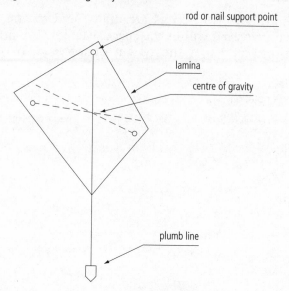

Centre of gravity of regular shaped bodies

For common regular shapes the centre of gravity is at the centre point. Figure 3.18 shows the centres of gravity of a rectangle, a square and a circle – all at the centre.

Figure 3.18 Centre of gravity of regular shaped bodies.

The centre of gravity of a triangle is at the intersection of the lines drawn from each vertex to the midpoint of the opposite side (Fig. 3.19).

Figure 3.19 Centre of gravity of a triangle.

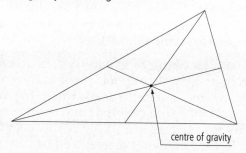

Stability

The position of the centre of gravity of a body affects its stability. Some things topple over quite easily, while others do not. This is due to the position of the centre of gravity. The design of tall vehicles such as buses, which can lean over when turning corners, must take into account the position of the centre of gravity. A bus has a low centre of gravity because the heavy mechanical components are close to the ground, which makes it very stable. Racing cars and go-carts have a very low centre of gravity, which enables them to negotiate corners at very high speeds. A body will topple when its weight (represented by the vertical line below its centre of gravity) acts outside its base, as illustrated in Fig. 3.20.

Figure 3.20 Stability.

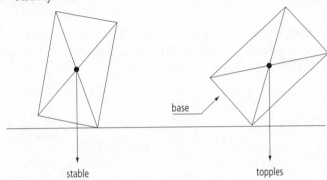

stable   topples

Stable, unstable and neutral equilibrium

A body is in stable equilibrium if, when it is slightly moved and then released, it returns to its original position. A body in unstable equilibrium will move away from its previous position when slightly displaced, while a body in neutral equilbrium will stay in its new position when slightly displaced (Fig. 3.21).

Figure 3.21 Stable, unstable and neutral equilibrium.

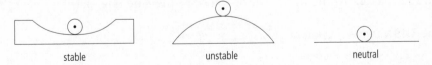

stable   unstable   neutral

Chapter review

Static system   a system which remains at rest when a force is applied

Force   that which changes the state of rest of a body on which it acts

Magnitude   the size

Direction   the line of action, i.e. horizontal, vertical or angled

Sense   the way the force acts on the direction, i.e. left, right, up, down

Point of application   where the force is applied

Scalar quantity   a quantity with magnitude (size) only, i.e. a temperature of 20°

Vector quantity   a quantity with magnitude and direction, i.e. a force of 10 kN acting on the left side of an object

Vector   a line which completely represents a quantity and has magnitude, direction, sense and a point of application

Resultant   a force which can be applied at a point and have the same effect as any number of forces acting at the same point

Equilibrant   the force required to keep the system static; it acts in the opposite direction to the resultant

Centre of gravity   the point through which the weight of an object acts

Exercise 3.1

Determine the magnitude and direction of the force required to maintain equilibrium for the systems (a) to (f) shown in Fig. 3.22. Give the direction clockwise from the nearest force.

Figure 3.22   Systems for Exercise 3.1.

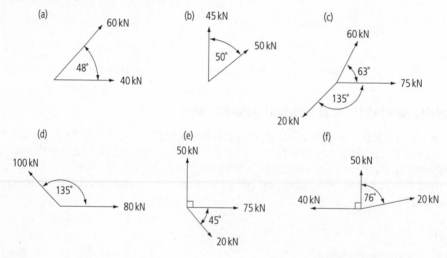

Activity sheet

Static systems

Investigate a static system by carrying out a force board experiment for three forces acting at a point.

List the apparatus used in the experiment.

Draw a space diagram of the system.

Draw the vectors end to end to a suitable scale and determine the resultant and equilibrant.

Prove your equilibrant value correct by placing a pulley and weight equal to the equilibrant value.

Determine the equilibrant for the systems shown in Fig. 3.23 graphically, and then prove them correct on a force board.

Figure 3.23

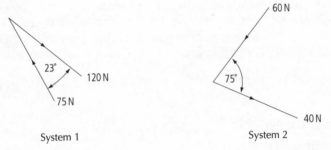

System 1   System 2

You can investigate further systems by carrying out the experiment with several forces acting at a point and using Bow's notation to determine the equilibrant. Construct these systems on the force board to prove your results.

○— **Key skill opportunities in application of number and information technology can be demonstrated in this activity and recorded as portfolio evidence.**

chapter

4

Pressure systems

4.1 Pressure

When a force acts on an object it can be concentrated at a point or spread over a larger area. For example, if you wear skis your weight is spread over a larger area than if you wear boots. In soft snow skis will keep you on the surface of the snow, when in boots you would sink in. Although the applied force is the same, the area has changed, and the pressure on the snow is greater with the boots than the skis.

Pressure is defined as the force acting on unit area, and is calculated using the formula

$$\text{pressure} = \frac{\text{force (N)}}{\text{area (m}^2)}$$

The pressure formula can be transposed in the following ways to determine force and area, thus

$$\text{force} = \text{pressure} \times \text{area} \quad \text{and} \quad \text{area} = \frac{\text{force}}{\text{pressure}}$$

The unit of pressure is the pascal (Pa), where $1\ \text{Pa} = 1\ \text{N/m}^2$. The unit N/mm^2 is sometimes used for calculations when the area is small, and the bar is also a useful unit. 1 bar is equal to 100 000 Pa or 10^5 Pa, and is the approximate value for atmospheric pressure.

4.2 Atmospheric pressure

Atmospheric pressure is the pressure due to the weight of air that surrounds us. The atmosphere exerts a pressure of approximately 101.3 kPa or 1.013 bar. Atmospheric pressure can be measured with an instrument called a barometer. Figure 4.1 shows a mercury barometer. Atmospheric pressure acts on the open surface of the mercury container, pushing the mercury up the vacuum tube.

Figure 4.1 A barometer.

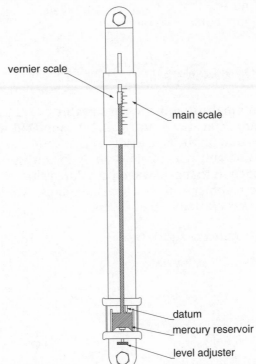

vernier scale

main scale

datum

mercury reservoir

level adjuster

Atmospheric pressure varies with the altitude, which means that the level of the mercury will change.

The weather is also affected by atmospheric pressure and a low barometer reading often indicates bad weather, while a high barometer reading indicates fine weather. Weather maps show lines, known as isobars, which join together places of equal pressure. When the isobars are close together the pressure gradient is large, which causes strong winds and storms.

To determine atmospheric pressure using a barometer, it is necessary to measure the height of the mercury column. This is done by resetting the level of mercury in the reservoir to the fixed pointer, which acts as a datum. Atmospheric pressure will normally support 760 mm of mercury in the tube. The following formula can be used to calculate atmospheric pressure:

pressure = $\rho g h$

where ρ (rho) is the density of the mercury in the tube (13 600 kg/m^3), g is the acceleration due to gravity (9.81 m/s^2) and h is the height of the mercury in the tube (760 mm = 0.76 m).

Thus

pressure = 13 600 × 9.81 × 0.76

= 101 396

= 101.4 kN/m^2 or kPa

Note: it is common to find that atmospheric pressure is stated as 760 mm mercury, which means that atmospheric pressure is equal to the pressure exerted by a 760 mm column of mercury.

4.3 Vacuum pressure

Pressures below normal atmospheric pressures are called vacuum pressures, or vacuums. The barometer shown in Fig. 4.2 is calibrated so that atmospheric pressure becomes zero on the gauge side. This is known as gauge pressure and is used in hydraulic and pneumatic systems for measuring pressure. If the gauge was calibrated so that zero was an absolute value, as shown on the left-hand side of the barometer, then a pressure gauge would never read zero, unless a vacuum was created in the system.

Figure 4.2 Absolute pressure and gauge pressure.

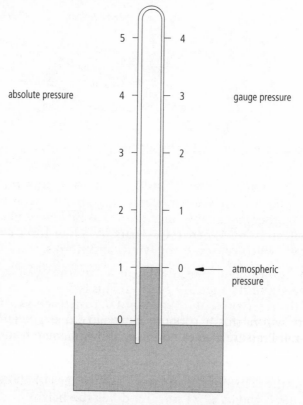

Absolute pressure readings are taken from the left-hand scale on the barometer. At absolute zero the mercury in the tube would disappear completely, since there would be no pressure to support the mercury column.

4.4 Pressure gauges

Engineering systems such as hydraulics or pneumatics use pressure gauges to indicate system pressure to machine operators and engineers. The most commonly used gauges in hydraulic systems are Bourdon tube gauges (Fig. 4.3).

Figure 4.3 A Bourdon tube pressure gauge.

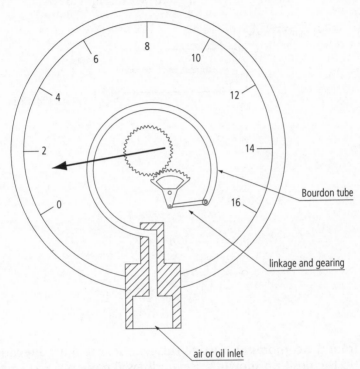

The gauge operates by the recoil of a flat tube, known as a Bourdon tube. Pressure from oil or compressed air in the system makes the tube flex, similar to a hosepipe lying on a lawn with the water turned on. The movement of the tube is transmitted through the gearing to the pointer, which shows the system pressure.

The recognized symbol for a pressure gauge in an engineering system is shown in Fig. 4.4. Pressure gauges normally show pressure readings in bar or kPa.

Figure 4.4 Symbol for a pressure gauge.

4.5 Pneumatic systems

In a pneumatic system, air under pressure is used to drive motors and cylinders to do useful work. The type of work output depends on the system, for example air motors can be used for winches to lift objects, and cylinders can be used on production lines to move parts from one area to another. An example of a basic pneumatic system is shown in Fig. 4.5.

Figure 4.5 A pneumatic system.

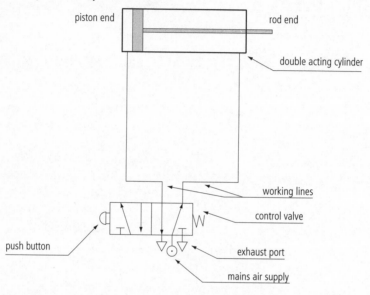

Mains air from the compressor is fed to the cylinder though a directional control valve. In the position shown, the cylinder will remain at rest since the air is directed to the rod end of the cylinder. To extend the piston, the directional control valve must be operated using the push button. The valve changes position and air is directed into the piston end of the cylinder. The pressure acts on the piston inside the cylinder, producing a force which will extend the piston. When the directional valve push button is released, the valve returns to its original position and the air is directed into the rod end, making the piston retract. This force is lower than the extending force since the cross-sectional area on the retract side is smaller because of the space taken up by the piston rod. A bicycle pump can be used to simulate a cylinder by blowing into the valve. Pressure acts on the piston inside the pump and the piston rod (handle) will extend.

4.6 Area of contact

The greater the area over which a force acts the lower the pressure will be. Consider a force of 200 N acting on the two areas shown in Fig. 4.6.

Figure 4.6 Area of contact.

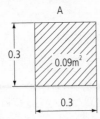

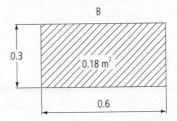

Area A $= 0.3 \times 0.3 = 0.09$ m^2

Pressure acting on A = $\dfrac{\text{Force}}{\text{Area}}$

Pressure acting on A = $\dfrac{200}{0.09}$

Pressure acting on A = $\quad$ 2222.2 N/m^2

Area B $= 0.6 \times 0.3 = 0.18$ m^2

Pressure acting on B = $\dfrac{\text{Force}}{\text{Area}}$

Pressure acting on B = $\dfrac{200}{0.18}$

Pressure acting on B = $\quad$ 1111.11 N/m^2

It can be seen that the pressure acting on the square surface area is greater than the pressure acting on the rectangular area, although the applied force is the same. This is because the area of the square surface is smaller than the area of the rectangular surface. In other words, the rectangular surface, being bigger, spreads the load out more.

EXAMPLE 4.1

Determine the pressure acting on a piston of cross-sectional area 50 mm^2 when a force of 200 N is applied.

Pressure $= \dfrac{\text{force}}{\text{area}}$

$= \dfrac{200}{50}$

$= 4$ N/mm^2

EXAMPLE 4.2

A mass of 20 kg is placed on a bench. If the mass has a base of 0.15 m $\times$ 0.25 m, determine the pressure acting on the bench.

Note that the mass (kg) must be converted to a force (N) using

Force $=$ mass $\times$ acceleration due to gravity

$= 20 \times 9.81$

$= 196.2$ N

Area to which the force is applied $= 0.15 \times 0.25 = 0.0375$ m^2.

$$\text{Pressure} = \frac{\text{force}}{\text{area}}$$

$$= \frac{196.2}{0.0375}$$

$$= 5232$$

Therefore the pressure on the bench due to the 20 kg mass is 5232 N/m^2 or 5.232 kPa.

EXAMPLE 4.3

A block of steel weighs 750 N and its base is a square of side 2 m. Determine the pressure the block exerts when placed on a floor.

$$\text{Pressure} = \frac{\text{force}}{\text{area}}$$

$$= \frac{750}{4}$$

$$= 187.5$$

Therefore the pressure on the floor due to the block of steel is 187.5 N/m^2.

EXAMPLE 4.4

A spring-operated non-return valve prevents air from entering a system until a pressure of 200 kPa acts on the valve seat. If the closing force by the spring is 40 N, determine the minimum diameter of the valve seat.

$$\text{Pressure} = \frac{\text{force}}{\text{area}} \quad \therefore \text{ area} = \frac{\text{force}}{\text{pressure}}$$

Note that the pressure = 200 kPa = 200 × 10^3 Pa or 200 000 Pa.

$$\text{Area} = \frac{40}{200\ 000} = 0.0002 \text{ m}^2$$

Since area $= \dfrac{\pi D^2}{4} = 0.0002$, then we can rearrange this formula to find diameter D.

$$D^2 = 4 \times \frac{\text{area}}{\pi}$$

$$D = \left(\frac{4 \times 0.0002}{\pi}\right)^{1/2} \text{(note that the power of ½ is the same as the } \sqrt{} \text{ [square root])}$$

$$D = 0.0159 \text{ m}$$

This is a small value when the metre unit is used, so it may be given in millimetres. Since there are 1000 mm in 1 m, we can convert to millimetres by multiplying by 1000, i.e. 0.0159 × 1000 = 15.9 mm.

$\therefore$ The minimum piston diameter required is 15.9 mm.

·EXAMPLE 4.5

A cylinder in a pneumatic system has a piston of diameter 20 mm. Determine the force exerted as the piston in the cylinder extends if the system air pressure is 0.6 N/mm^2.

Force = pressure × area

$$= 0.6 \times \frac{\pi \times 20^2}{4}$$

$$= 0.6 \times 314.16$$

$$= 188.49$$

Force exerted on the extending piston = 188.5 N.

EXAMPLE 4.6

Determine the force in the cylinder shown in Fig. 4.7 (a) when the piston extends and (b) when the piston retracts, given that the air pressure supply is 0.85 N/mm^2.

Figure 4.7 Pneumatic cylinder. Diameter of piston, 30 mm; diameter of piston rod, 20 mm.

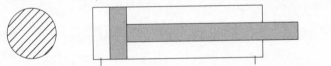

Force when piston extends = pressure × area of cylinder

$$= 0.85 \times \frac{\pi \times 30^2}{4}$$

$$= 0.85 \times 707$$

$$= 601 \text{ N}$$

Force when piston retracts = pressure × (area of cylinder − area of piston rod)

The area can be calculated using

$$\frac{\pi}{4} \times (D^2 - d^2)$$

where D is the diameter of the piston and d is the diameter of the piston rod,

$$\therefore \text{force} = 0.85 \times \frac{\pi}{4} \times (30^2 - 20^2)$$

$$= 0.85 \times 392.86$$

$$= 334 \text{ N}$$

Thus (a) the force acting on the piston as it extends is 601 N, and (b) the force acting on the piston as it retracts is 334 N.

4.7 Pressure in liquids

We have already seen that liquids also exert pressure. If the density or depth of a liquid increases, the pressure will increase. The relationship between pressure, depth and density is given by

Pressure $= \rho g h$

Where ρ (rho) is the density of the liquid in kg/m^3, g is the acceleration due to gravity, 9.81 m/s^2, and h is the depth of the liquid in metres, m.

It should be remembered that the greater the depth, the larger the pressure, and the greater the liquid density, the larger the pressure.

A tank containing water of density 1000 kg/m^3 is shown in Fig. 4.8. The pressure at the base of the tank can be calculated using $p = \rho g h$.

Figure 4.8 Pressure in a water tank.

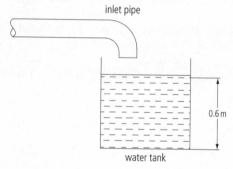

For the tank $h = 0.6$ m

$\therefore p = \rho g h$

$= 1000 \times 9.81 \times 0.6$

$= 5886$

Hence the pressure at the base of the tank is 5886 N/m^2 or 5886 Pa.

EXAMPLE 4.7

A rain water barrel 1 m high is approximately half full. Determine the pressure at the bottom of the barrel if the density of water is 1000 kg/m^3.

$p = \rho g h$

$= 1000 \times 9.81 \times 0.5$

$= 4905$

The pressure at the bottom of the barrel is 4905 N/m^2 or 4905 Pa.

If the diameter of the barrel is 0.6 m, determine the force acting on the bottom of the barrel.

Force = pressure × area

$$= 4905 \times \frac{\pi \times 0.6^2}{4}$$

$$= 4905 \times 0.283$$

$$= 1387$$

The force acting on the bottom of the barrel is 1387 N.

EXAMPLE 4.8

A water tank in a house loft is 2.3 m above a showerhead in an upstairs bathroom, and 5 m above a tap in the kitchen. Determine the pressure at the shower and at the kitchen tap.

Pressure at the shower $= \rho gh$, where $h = 2.3$ m.

$$p = 1000 \times 9.81 \times 2.3$$

$$= 22\,563 \text{ N/m}^2$$

$$= 22.563 \text{ kN/m}^2 \text{ or } 22.563 \text{ kPa}$$

Pressure at the tap $= \rho gh$, where $h = 5$ m.

$$p = 1000 \times 9.81 \times 5$$

$$= 49\,050 \text{ N/m}^2$$

$$= 40.05 \text{ kPa or } 40.05 \text{ kN/m}^2$$

4.8 The U-tube manometer

The U-tube manometer is a much simpler pressure-measuring instrument than the pressure gauge. It consists of a U-shaped glass tube open to the atmosphere at one end and connected to a pressure source at the other, for example a household gas supply. When the gas supply is turned on, gas pressure pushes the liquid up the open side of the U-tube. When the level settles the difference in height of the liquid in the tube can be measured (Fig. 4.9). This can be used to determine the difference in pressure between the two points.

The formula pressure $= \rho gh$ can be used to determine the difference in pressure between the two points. To determine the gas pressure the value should be added to atmospheric pressure, i.e.

gas pressure = pressure difference + atmospheric pressure

In this case the gas pressure is an absolute value since atmospheric pressure is added to the pressure difference.

Figure 4.9 A manometer.

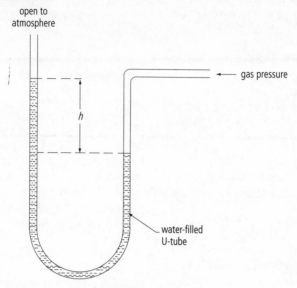

EXAMPLE 4.9

Determine the pressure difference in a U-tube manometer connected to a gas supply if the difference in water levels in the tube is 180 mm. Find the pressure in the gas (absolute) to produce the difference in water levels taking atmospheric pressure as 103.25 kN/m².

Note that the height of liquid must be in metres, hence 180 mm = 0.18 m.

$$\text{Pressure} = \rho gh$$
$$= 1000 \times 9.81 \times 0.18$$
$$= 1765.8 \text{ N/m}^2$$

The difference in pressure is 1765.8 N/m² = 1.77 kN/m².

Absolute pressure of the gas = pressure difference + atmospheric pressure

$$= 1.77 + 103.25$$
$$= 105.02 \text{ kN/m}^2$$

The absolute pressure needed to produce a 180 mm difference in water levels is 105.02 kN/m².

EXAMPLE 4.10

A U-tube manometer contains mercury of density 13 600 kg/m³. One leg is open to the atmosphere and the other is connected to a vessel containing gas (Fig. 4.10). Calculate the pressure difference which produces a height difference in the two legs of 30 mm.

$$\text{Pressure} = \rho gh$$
$$= 13\,600 \times 9.81 \times 0.03$$
$$= 4002.48 \text{ N/m}^2$$

The pressure difference is 4.002 kN/m².

Figure 4.10 Mercury manometer.

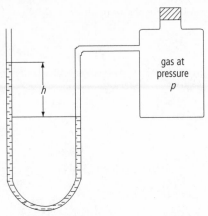

4.9 The hydraulic jack

Pressure in a system can be used to produce large forces to do work. In a hydraulic jack, oil is used to transmit pressure. Like other liquids, oil is virtually incompressible, so any increase in pressure produces a force on the area in contact with the oil. Hydraulic jacks are used in garages to lift cars and the principle of their operation is known as force multiplication, where a small input force on the handle is transmitted to a large lifting force at the load.

Figure 4.11 shows how a hydraulic jack works. When the pump lever is operated, oil is pushed down the small cylinder by the pumping piston. The oil is forced out of the cylinder, along a pipe and into a wider cylinder, which houses the lifting piston. The pressure in the oil forces the lifting piston upward, raising the load a little at a time. When the pumping piston is moved back up the cylinder a non-return valve keeps the oil in the larger cylinder. As the piston rises in the pumping cylinder, atmospheric pressure acting on the surface of the reservoir causes a second non-return valve at the base of the reservoir to open. Oil is forced into the pumping cylinder, filling it up, ready for the next downward stroke. The non-return valve at the base of the reservoir closes as the piston is moved down, and the cycle is repeated, until the load is lifted to the desired height.

When the job is complete and the load requires lowering, a lowering valve is opened which allows the load to descend, forcing oil to flow back to the reservoir.

Figure 4.11 Hydraulic jack: (a) lowering valve, (b) and (c) non-return valves, (d) oil return line, (e) oil reservoir, (f) pump lever, (g) pumping piston, (h) lifting piston.

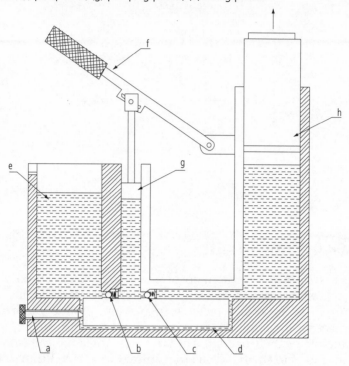

4.10 Force multiplication

Hydraulic jacks follow a law known as Pascal's law. This states that pressure applied to a static and confined liquid is transmitted undiminished in all directions and acts with equal force on equal areas and at right angles to them (Fig. 4.12) This means that when a force is applied to one end of the column of liquid, it is transmitted through the column to the other end, but also to the sides of the vessel without losses.

If we apply a force of 100 N to the small piston of a hydraulic jack, then a pressure is created since pressure = force per unit area. The pressure at the large piston will be equal to that in the smaller one (Pascal's law), but the force or load that can be lifted will be much greater than the input force (Fig. 4.13).

Let the area of piston A be 50 mm^2 and the area of piston B be 1000 mm^2. Then

Pressure at A = $\dfrac{\text{force}}{\text{area}}$ = $\dfrac{100}{50}$ = 2 N/mm^2

Pressure at B = pressure at A = 2 N/mm^2

Force (load) at B = pressure × area B,

∴ force = 2 × 1000

= 2000 N

Figure 4.12 Pascal's law.

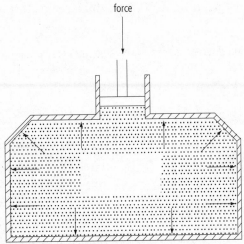

Figure 4.13 Force multiplication in a hydraulic jack.

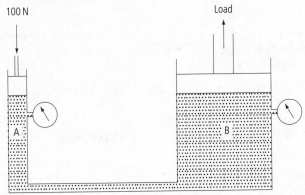

So a load of 2000 N can be lifted at B by an effort of only 100 N at A.

This value could also be calculated using the ratios of the areas of the two pistons:

$$\text{area ratio} = \frac{\text{area of piston A}}{\text{area of piston B}} = \frac{50}{1000} = 1:20$$

Since the area ratio is 1:20, then the effort is multiplied by the ratio at the load, i.e.

$$\text{load that can be raised} = \text{input} \times \text{area ratio}$$

$$= 100 \times 20$$

$$= 2000 \text{ N}$$

It may appear that we are converting a small effort into lifting a large load, without loss. However, the law relating to the conservation of energy states that energy can neither be created nor destroyed. In this instance we are increasing the load that can be lifted, but the distance is sacrificed as the lifting piston only moves a very short distance.

The distance moved can also be calculated by the area ratio method (Fig. 4.14). If the small piston moves 200 mm, the large piston will move 1/20th of this distance.

Figure 4.14   Calculating distance moved by the area ratio method.

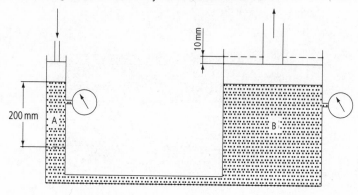

$$\text{distance moved} = 200 \times \frac{1}{20} = 10 \text{ mm}$$

This can also be proved by working out the volume of oil pumped from the small cylinder to the large cylinder, where the stroke of the small cylinder represents the length, i.e.

$$\text{Volume pumped from small cylinder} = \text{area} \times \text{length}$$

$$= 50 \times 200$$

$$= 10\,000 \text{ mm}^3$$

This volume is delivered to the large cylinder, where the length is equal to the distance the piston will rise, i.e.

$$\text{volume} = \text{area} \times \text{length}$$

then

$$\text{length} = \frac{\text{volume}}{\text{area}}$$

$$\text{length} = \frac{10\,000}{1000}$$

$$\text{length} = 10$$

∴ the piston will move a distance of 10 mm, raising the load.

EXAMPLE 4.11

The hydraulic jack shown in Fig. 4.15 has an area ratio of 1:100. Determine the load that can be raised by an effort of 100 N on the pumping piston. If the pumping piston stroke is 100 mm, calculate the distance the load will rise.

Since there is an area ratio of 1:100, the load that can be lifted will be 100 times greater than the input force owing to force multiplication. Thus

Figure 4.15 Calculating load and distance raised by the area ratio method.

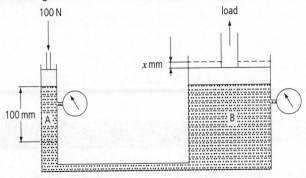

Load that can be lifted = input × 100

= 100 × 100

= 10 000 N

The distance moved will be 1/100th of the distance moved by the pumping piston,

∴ Distance moved = distance moved by pumping piston × 0.01

= 100 × 0.01

= 1 mm

EXAMPLE 4.12

For the hydraulic jack shown in Fig. 4.16, determine the force required and the length of stroke of the pumping piston to raise a load of 20 kN a distance of 4 mm in one stroke.

Figure 4.16 Calculating force and length of stroke of the pumping piston in a hydraulic jack. Area of piston A, 30 mm^2, area of piston B, 2400 mm^2.

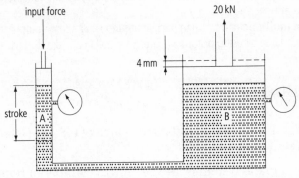

For this hydraulic jack

$$\text{area ratio} = \frac{\text{area A}}{\text{area B}} = \frac{30}{2400} = 1{:}80$$

∴ force required on the pumping piston $= \frac{20000}{80} = 250$

The input force to lift 20 000 N is 250 N, since the effort will be multiplied 80 times.

The stroke of the pumping cylinder is equal to the distance the load rises multiplied by the area ratio.

Stroke = 4 × 80

= 320

Hence the stroke of the pumping cylinder will be 320 mm.

Chapter review

Pressure   pressure is defined as the force acting on unit area, and is calculated using the formula

$$\text{pressure} = \frac{\text{force (N)}}{\text{area (m}^2)}$$

The SI unit of pressure is the pascal (Pa), where 1 Pa = 1 N/m^2

Atmospheric pressure   the pressure due to the weight of air that surrounds us

Absolute pressure   pressure readings that use absolute zero as the starting point

Mercury barometer   a vacuum tube filled with mercury, which rises up the tube depending on the atmospheric pressure. The height of mercury can be used to determine daily atmospheric pressure changes.

Manometer   a U-tube used to measure the difference in pressure between two points, given by

$$\text{pressure} = \rho g h$$

where ρ (rho) is the density of the liquid (kg/m^3), g is the acceleration due to gravity (9.81 m/s^2) and h is the depth of the liquid (in m)

Pascal's law   states that pressure applied to a static and confined liquid is transmitted undiminished in all directions and acts with equal force on equal areas and at right angles to them

Force multiplication   the principle of producing a large output force from a small input force by using pistons of different areas in a hydraulic jack

Exercise 4.1

1 A force of 20 kN acts on a steel plate of area of 0.5 m^2. Determine the pressure acting on the plate.

2 Air in a pneumatic system is directed into a cylinder of bore 20 mm. If the air pressure is 0.6 N/mm^2, determine the force acting on the piston (note that piston diameter = bore).

3 Determine the force in the cylinder shown in Fig. 4.17 (a) for the cylinder extending (b) for the cylinder retracting, given that the air pressure supply is 0.75 N/mm^2.

Figure 4.17 Pneumatic cylinder. Diameter of piston, 30 mm; diameter of piston rod, 20 mm.

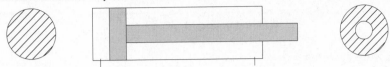

4 A water-filled U-tube manometer is connected to a gas supply producing a height difference in the legs of the tube of 120 mm. Determine the difference in pressure in the legs.

5 A vessel containing gas is connected to a water-filled U-tube manometer and produces a height difference in the legs of 80 mm. Determine the pressure of the gas in the vessel.

6 A mercury barometer reading shows 755 mm of mercury. If the density of mercury is 13 600 kg/m^3, determine the value for atmospheric pressure.

7 A force of 10 kN is applied to a hydraulic jack with an area ratio of 1:150. Determine the load that can be raised and the pressure in the system if the diameter of the small cylinder is 20 mm.

8 The pumping piston of a hydraulic jack has a stroke of 150 mm. If the area ratio for the cylinders is 1:100, determine the distance a supported load will rise.

9 For a hydraulic jack with an area ratio of 1:200, determine the load that can be raised and the distance the load will rise if the pumping piston travels 120 mm when a 150 N force is applied.

10 A hydraulic jack lifts a load of 200 kN. If the area ratio for the cylinders is 1:100, determine the force on the pumping piston.

11 In a hydraulic press, the pumping piston has a cross-sectional area of 200 mm^2 and the lifting cylinder piston has a cross-sectional area of 2400 mm^2. Calculate the force on the pumping piston if the press produces a force of 2000 N.

12 A hydraulic jack has pumping and lifting pistons of diameters 30 mm and 450 mm, respectively. If a force of 150 N is applied to the jack, determine the load that can be lifted.

Activity sheet

Pressure systems

Task 1

Conduct an experiment to determine the supply pressure of gas using a water-filled manometer. You will need a manometer, a gas supply, water, rubber hose and a rule. A calculator will be useful. The day's value for atmospheric pressure will also be required. This can be obtained from a barometer reading or you can take the standard value of 101.325 kPa.

Carry out the experiment as follows.

- Connect the apparatus as shown in Figure 4.9 and check the connection between the gas supply and the manometer.

- Turn on the gas supply slowly and observe the water level rise in the manometer.

- When the water levels settle, measure the difference in water levels of the U-tube.

- Determine the pressure difference in levels using the formula pressure $= \rho g h$.

- Determine the gas pressure by adding the pressure difference to the day's atmospheric pressure reading.

Task 2

Using a barometer determine the value of atmospheric pressure over five consecutive days. Construct a graph to display the differing daily pressures and compare the values with the national figures as supplied on the daily weather forecasts.

> **○━ Key skill opportunity – numeracy, communication and information technology skills can be demonstrated with this activity. The daily calculations of atmospheric pressure can be collated and presented in various ways which can be used as development evidence.**

chapter

5 Elasticity

When a force is applied to a spring, its length changes. The length will increase if the spring is stretched, while it will decrease if the spring is compressed. Many examples of engineering systems use this relationship, which is of great importance.

Springs are used in many everyday systems, some obvious, others not so. For example:

- vehicle suspension systems use coil and leaf springs to dampen jolting due to bumps in roads
- trains use springs under carriages to provide a more comfortable ride
- bicycle brakes use springs to return the brake lever
- coil springs are used in door handles and latches
- small springs are used under audio equipment function selectors

There are many other applications of systems using springs, large and small, and springs made from different materials. Steel, plastic and rubber all have elasticity, and it is this property which enables them to be used as springs.

5.1 What is meant by elasticity?

A material which returns to its original shape upon the removal of a force is said to be an elastic material. Steel, brass, copper, aluminium and rubber are examples of materials with elastic properties.

A rubber band will increase in length if it is pulled and return to its original length when the pulling force is removed (Fig. 5.1).

Fig. 5.1 Stretching a rubber band.

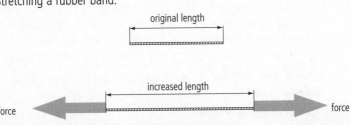

A spring will reduce in length if compressed (squeezed), and return to its original length when the compression force is removed (Fig. 5.2).

Fig. 5.2   Compressing a spring.

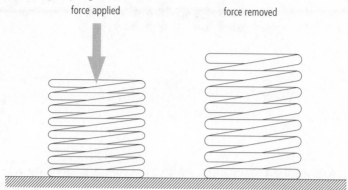

Some materials do not return to their original shape when the applied force is removed. These materials are known as plastic. Lead and plasticine are examples of materials with plastic properties.

5.2   The spring balance

We can use an elastic material to make a spring, for use in a spring balance. The spring balance is a basic hand-held measuring instrument used to weigh objects. The spring balance is often referred to as a **newton meter**. Anglers use a spring balance to determine the weight of their catch. Figure 5.3 shows a typical spring balance.

The balance works by recording an applied load (weight) on a scale giving a direct readout of the weight. The pointer moves as the spring extends due to the load. When the load is removed the spring returns to its original size and the scale reads zero.

5.3   The relationship between applied force and change in length of an elastic material

When a load or force is applied to an elastic material the material will increase or decrease in length, as shown in Figs 5.1 and 5.2. Further loads will cause the material to change length again. The change in length will vary for different materials, depending on the stiffness of the material. A rubber band would stretch more than a steel spring as steel is a stiffer material. Figure 5.4 shows the change in length of a spring as greater loads are applied.

An experiment may be performed in a science laboratory or workshop to determine the relationship between applied force and change in length using the equipment shown in Fig 5.5.

Fig. 5.3 A spring balance.

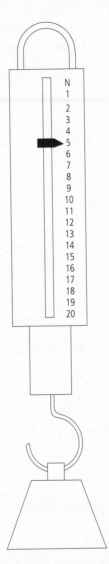

The results of this type of experiment can be recorded and a graph plotted of the applied force against the resulting extension (Fig. 5.6).

A straight line can be drawn through the plotted values of applied load and extension, showing that the extension is proportional to the applied force. This is known as **Hooke's law**, and applies to all elastic materials up to a point known as the elastic limit. After this point the resulting extension will no longer be proportional to the applied force and the material will be permanently deformed if further loads are applied.

The graph can be used to determine intermediate values which were not found in the experiment.

From the graph we can see that a force of 45 N produces an extension of 6.75 mm and an extension of 5 mm is the result of a 33 N force.

Fig. 5.4 Increase in length of a spring with increase in load.

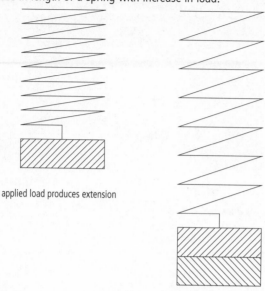

applied load produces extension

greater load produces larger extension

Fig. 5.5 Determining the relationship between extension and load.

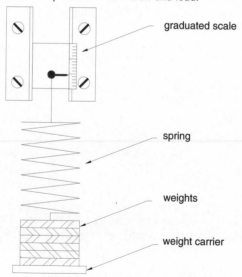

graduated scale

spring

weights

weight carrier

5.4 The slope or gradient of a force–extension graph

The straight line produced by a force–extension graph has a particular slope or gradient. The value of the gradient is called the stiffness – it tells us how much force is needed to produce a unit increase in length. The stiffness is usually given in units of N/mm. The following example shows two methods of finding the stiffness.

Fig. 5.6 Force–extension graph.

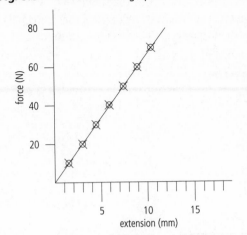

Force (N)	Extension (mm)
0	0
10	1.5
20	3.0
30	4.4
40	5.9
50	7.5
60	9.0
70	10.5
80	12.0

Fig. 5.7 Using the force–extension graph.

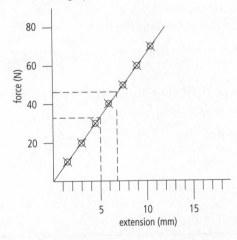

EXAMPLE 5.1

The table shows the extension produced by hanging weights on a spring balance.

Weight (N)	Extension (mm)
0	0
50	10
75	15
100	20
150	30
200	40

To determine the stiffness of the material, we can use either of two methods

(a) Plot the graph of force (weight) against extension and determine the slope by measuring. The slope is the difference in the *y* values between two points on the line divided by the difference in their *x* values (see Chapter 2).

(b) Pick two sets of values and apply the following method

Values selected: 50 N, 10 mm and 150 N, 30 mm

Change in force = 150 N − 50 N = 100 N

Change in length = 30 mm − 10 mm = 20 mm

Divide the force value by the length

100 ÷ 20 = 5

The spring stiffness is therefore 5 N/mm.

When the value for the material stiffness is known, or calculated as in Example 5.1, it can be used to determine the resulting extension if the force is known, or the applied force if the extension is known.

EXAMPLE 5.2

A wire extends 0.2 mm when a force of 25 N is applied. Find the stiffness of the wire in N/mm.

Force ÷ extension = constant = stiffness of material

force = 25 N

extension = 0.2 mm

thus constant = 25 ÷ 0.2 = 125 N/mm

The stiffness of the wire is 125 N/mm.

EXAMPLE 5.3

A wire extends 0.5 mm when a force of 250 N is applied. Find the force required to produce an extension of 1.25 mm.

The stiffness *k* can be found using the formula

$$\frac{force}{extension} = \text{constant } k$$

$$\frac{250}{0.5} = 500 = k$$

Rearranging the formula to find force:

force = $k \times$ extension

then force = 1.25 × 500 = 625

The force required to extend the wire 1.25 mm is 625 N.

EXAMPLE 5.4

A spring extends by 0.3 mm when a load of 1.5 kg is applied. Determine the stiffness of the spring.

Spring stiffness = force ÷ extension

Force = mass × gravity   (where gravitational constant = 9.81 m/s²)

Force = 1.5 × 9.81 = 14.715 N

Spring stiffness = 14.715 ÷ 0.3 = 49.05

The spring stiffness is 49.05 N/mm.

Chapter review

Elasticity   the ability of a material to return to its original shape and size on removal of external forces

Plastic (plasticity)   the property of a material which remains permanently deformed on removal of an external force

Elastic limit   the point at which a material no longer obeys Hooke's law, i.e. where the extension will no longer be proportional to the applied force

Slope or gradient   found by dividing the force by the extension; this represents the stiffness of the material

Material stiffness   a rubber band stretches more easily than a steel spring because steel is a stiffer material

Exercise 5.1

1 (a) A mass of 20 kg extends the spring on a spring balance by 25 mm. Determine the spring stiffness.

(b) What mass would extend the same spring 45 mm?

2 The table shows the extension of a spring in a spring balance when different weights are applied.

Weight (N)	0	2	4	6	8	10	12
Extension (mm)	0	6	12	17.5	24.6	30.5	37

Plot the graph of force against extension and determine the stiffness of the spring.

3 Define what is meant by an elastic material and state Hooke's law.

4 The length of a spring is 0.6 m when it is subjected to a 5 N force, and 1.2 m when it is subjected to 10 N. What would the spring's length be if a 7.5 N force is applied, and the spring obeys Hooke's law?

5 A load of 50 N produces a 0.2 mm extension on a wire. If the wire obeys Hooke's law, calculate:

(a) the extension a 125 N load produces;

(b) the load required to produce a 2 mm extension.

6 A spring 100 mm long extends to 125 mm long when a load of 40 N is applied. Determine the load required to extend the spring to 150 mm.

Activity sheet

The spring balance

Conduct an experiment using a spring balance and weights to investigate Hooke's law. You will need the following equipment:

- a newton or gram spring balance

- a weight carrier and some suitable weights, 50 g would be ideal

- a stand to hang the balance on

- a rule, graph paper, pencil and a calculator might be useful

Draw a labelled diagram of the equipment set up and describe in detail the method used in this experiment.

Complete a results table similar to the one on page 61.

Draw the graph of force against extension from your results and calculate the slope of the graph.

Conclusion: answer the following

State the stiffness of the spring material in newtons per millimetre (N/mm) and comment on the quality of the graph.

Has Hooke's law been proved?

Describe how the graph can be used to estimate extension for a known force value.

> o—¬ **Key skill opportunity – the data from your equipment can be used as evidence of application of number skills. It may also be fed into a spreadsheet/graphics package and a graph can be produced by the computer.**

6

The moment of a force

6.1 Turning moment

Figure 6.1 shows a spanner being used to tighten a nut and bolt on an assembly. A force is applied on the spanner body. This turns the spanner in the direction of the applied force, causing the nut to tighten.

Fig. 6.1 Tightening a nut and bolt assembly with a spanner.

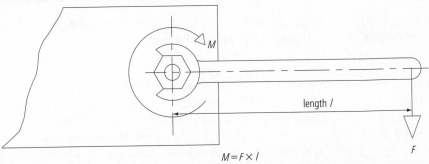

$$M = F \times l$$

The force applied to the nut at the perpendicular distance produces an effect known as a turning moment. Hence

turning moment = force × perpendicular distance

In the SI system of units force is measured in newtons (N) and distance in metres (m), so the turning moment is measured in newton metres (N m).

The term moment is acceptable instead of stating turning moment and N mm may be used in addition to N m if the perpendicular distance is small.

Other examples of turning moments are:

- the opening of a door by a lever or handle
- turning on a bath tap
- tightening a screw with a screwdriver

EXAMPLE 6.1

A spanner 300 mm long is used to tighten a bolt on a machine guard (Fig. 6.2). If the applied force is 150 N, determine the moment on the bolt.

Fig. 6.2   A spanner 300 mm long used to tighten a bolt with applied force 150 N.

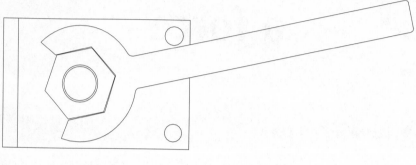

Moment = force × distance

= 150 N x 0.3m

= 50 Nm

The term **torque** is often applied to this type of situation, where the torque is the product of the force and the radius. Hence

torque = force × radius

Torque is a term widely used in engineering. Nuts, bolts and socket screws are tensioned to a predetermined level using a tool called a torque wrench, which eliminates the risk of overtensioning components. Wheel bearings and cylinder head fastenings on cars are examples of applications involving torque wrenches. It is important always to use a torque wrench when tensioning a component if the magnitude of the torque is specified in the installation or maintenance procedure.

EXAMPLE 6.2

Calculate the force required at a distance of 250 mm from the centre of a bolt if the resulting torque is 50 N m.

Torque = force × radius

Rearranging the formula:

force = torque ÷ radius

$$force = \frac{50}{0.25} = 200$$

i.e. a force of 200 N acting at a radius of 250 mm produces a torque of 50 N m.

6.2 Equilibrium

This term means 'at rest' and is often defined thus:

A body will remain in a state of rest or equilibrium provided that there are no external forces applied.

For example, a stationary object will remain so unless a force is applied to the object which moves it in the direction of the applied force.

6.3 The principle of moments

We have considered the turning effect produced by a force acting at a distance from a point. This turning moment affects the state of equilibrium of the object; however, the body can maintain equilibrium if the following two conditions are satisfied.

1 The resultant of the forces acting on the body is zero.
2 The sum of the clockwise moments about any point equals the sum of the anticlockwise moments about the same point.

Figure 6.3 shows an example of these conditions.

Fig. 6.3 A beam in equilibrium.

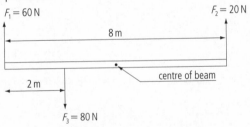

The resultant force = upward forces − downward forces

$$= 60 \text{ N} + 20 \text{ N} - 80 \text{ N}$$

$$= 0$$

This can also be proved by using the formula

sum of upward forces = sum of downward forces

$$60 + 20 = 80 \text{ N}$$

Method to show that the clockwise moments are equal to the anticlockwise moments

A clockwise moment will try to turn the beam clockwise and an anticlockwise moment will try to turn the beam anticlockwise.

We can take moments about any point on the beam, for example at the beam centre.

Clockwise moments = anticlockwise moments

$(4 \times 60) = (2 \times 80) + (4 \times 20)$

$240 = 160 + 80$

$240 \text{ N m} = 240 \text{ N m}$

We can check the above rule by taking moments at a different point, for example the right-hand end of the beam.

Clockwise moments = anticlockwise moments

$(8 \times 60) = (6 \times 80)$

$240 \text{ N m} = 240 \text{ N m}$

It should be noted that the 20 N force produces no turning moment at this point, since there is no distance involved.

EXAMPLE 6.3

Determine the force F required for the pivoted beam shown in Fig. 6.4 to maintain equilibrium.

Fig. 6.4   Pivoted beam in equilibrium.

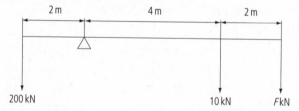

2 m   4 m   2 m

200 kN   10 kN   FkN

Taking moments at the pivot, clockwise moments = anticlockwise moments:

$(4 \times 10) + (6 \times F) = (2 \times 200)$

$40 + 6F = 400$

$6F = 400 - 40$

$6F = 360$

$F = \dfrac{360}{6}$

$F = 60 \text{ N}$

i.e. the force F required to maintain equilibrium is 60 kN.

Check answer using clockwise moments = anticlockwise moments:

$(4 \times 10) + (6 \times 60) = (2 \times 200)$

$40 + 360 = 400$

$400 \text{ N m} = 400 \text{ N m}$

EXAMPLE 6.4

A car suspension is shown in Fig. 6.5. If the force due to the road on the wheel is 750 kN find the force the shock absorber applies to maintain equilibrium.

Fig. 6.5   Car suspension.

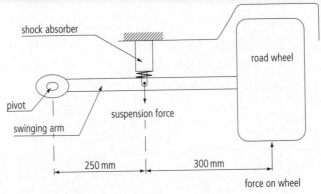

(Note: let F = force applied by shock absorber.)

Taking moments at the pivot, clockwise moments = anticlockwise moments:

$(0.25 \times F) = (0.55 \times 750)$

$0.25\, F = 412.5$

$F = \dfrac{412.5}{0.25}$

$F = 1650\,\text{N}$

Hence the force required by the shock absorber to keep the wheel on the road is 1650 kN.

The principle of moments can be used to solve a wide range of engineering problems, from how to build strong structures to how to mount overhead cranes in workshops and how best to support components being machined in a lathe.

Practical solutions to these problems have been developed:

- girders or pillars in a building support the weight of the floor above
- I-section girders are used to support and carry workshop overhead cranes
- bearings in a lathe chuck and tailstock support the workpiece and react to the force applied by the cutting tool

It should be noted that the forces at the points of support in the examples above are normally called reactions.

EXAMPLE 6.5

Determine the distance x mm required to maintain equilibrium in the system shown in Fig. 6.6.

Fig. 6.6 Pivoted system in equilibrium.

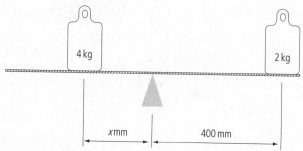

For equilibrium clockwise moments = anticlockwise moments:

$$400 \times 2 = 4x$$
$$800 = 4x$$
$$x = 800 \div = 200$$

The distance x will be 200 mm.

EXAMPLE 6.6

Figure 6.7 shows a crowbar being used to lift a crate of mass 500 kg. Determine the force required to lift the crate.

Fig. 6.7 Levering with a crowbar.

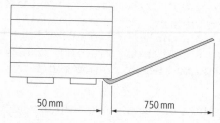

$$0.75 \times F = (500 \times 9.81) \times 0.05$$
$$0.75F = 245.25$$
$$F = \frac{245.25}{0.75} = 327 \text{ N}$$

The force required to lift the crate is 327 N.

EXAMPLE 6.7

Determine the reactions at the supports for the beam shown in Fig. 6.8. Prove your answer using the principle of moments.

Fig. 6.8   Reactions at supports using the principle of moments.

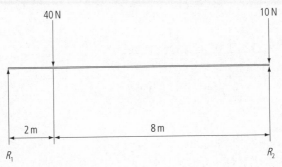

Taking moments about R_1 to find R_2:

clockwise moments = anticlockwise moments

$(40 \times 2) \times (10 \times 10) = 10R_2$

$80 + 100 = 10R_2$

$180 = 10R_2$

$R_2 = \dfrac{180}{10} = 18$

$R_2 = 18\,\text{N}$

Taking moments about R_2 to find R_1

clockwise moments = anticlockwise moments

$10\,R_1 = (40 \times 8)$

Note that the 10 N force is not considered since there is no distance involved this time.

$10\,R_1 = 320$

$R_1 = \dfrac{320}{10} = 32$

$R_1 = 32\,\text{N}$

Checking answers with the principle of moments:

sum of the upward forces = sum of the downward forces

$32 + 18 = 10 + 40$

$50 = 50$

∴ equilibrium will be maintained in the beam.

EXAMPLE 6.8

Determine the value of F needed to maintain equilibrium in the balance shown in Fig. 6.9 and the reaction R at the pivot.

Fig. 6.9 Balance in equilibrium.

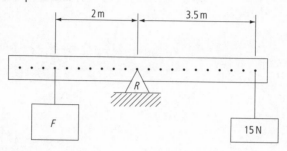

Taking moments about the pivot:

  clockwise moments = anticlockwise moments

  $15 \times 3.5 = 2 \times F$

  $52.5 = 2F$

  $F = \dfrac{52.5}{2}$

  $F = 26.25\,\text{N}$

The force required to maintain equilibrium is 26.25 N.

To find the reaction R at the pivot:

  sum of upward forces = sum of the downward forces

Here the reaction R is the upward force,

  $\therefore R$ = sum of the downward forces

  $R = 26.25 + 15$

  $R = 41.25\,\text{N}$

The reaction R at the pivot is 41.25 N.

EXAMPLE 6.9

A tyre lever 250 mm long is inserted into a wheel rim. A force of 100 N is applied at the end of the lever which pivots 50 mm from the end. Determine the force exerted on the tyre.

A simplified diagram of the lever is shown in Fig. 6.10, with the fulcrum representing the pivot point on the wheel rim, and F representing the force on the tyre.

  Clockwise moments = anticlockwise moments

In this case the 100 N force produces the clockwise moment, and the force on the tyre produces the anticlockwise moment.

Fig. 6.10   Using a tyre lever.

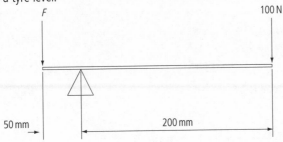

$\therefore 100 \times 200 = 50 \times F$

$20\,000 = 50F$

$\therefore F = \dfrac{20\,000}{50}$

$F = 400\,\text{N}$

The force applied to the tyre is 400 N.

EXAMPLE 6.10

A mechanic uses a spanner 300 mm long to tighten a cover plate on an engine. If the applied force is 80 N, determine the torque.

Torque = force × radius

$\qquad$ = 80 × 0.3

$\qquad$ = 24 N m

The torque applied is 24 N m.

Chapter review

Moment   the product of a force applied at a perpendicular distance

Torque   a force acting at a radius producing a turning effect

Equilibrium   a stationary object will remain in equilibrium unless a force is applied to the object which moves it in the direction of the applied force

Principle of moments   about any point:

sum of clockwise moments = sum of the anticlockwise moments

sum of the upward forces = sum of the downward forces

Exercise 6.1

1 Determine the moment(s) about the pivot for the situations shown in Fig. 6.11.

Fig. 6.11 Moments about different pivot systems.

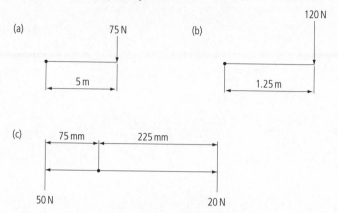

2 A force is applied to a spanner 300 mm long, which produces a torque of 45 N m. Determine the value of the force.

3 A force of 60 N acts on a pulley of 1.2 m diameter. Determine the torque acting on the pulley.

4 A turning moment of 600 N m is produced by a force of 150 N. Determine the perpendicular distance at which the force is applied.

5 A beam 6 m long is simply supported at its ends. If a 20 kN force is applied 2m from the left end of the beam, determine the reactions to the load. Prove your answers using the principle of moments.

Activity sheet

The moment of a force

Carry out an investigation of moments using the following equipment:

- a metre rule (beam), drilled at equal intervals along its length
- two spring balances
- two retort stands
- weights of varying sizes

Set the equipment up as shown in Fig. 6.12, and suspend a weight at one position of the beam. Place a second larger weight at a second distance along the beam.

Record the readings of the spring balances and enter in a table of results.

Apply the principle of moments and calculate the value of the reactions; enter these in the table.

Compare the reaction values with the spring balance readings and comment on the results.

Repeat the exercise for additional weights across the beam.

Fig. 6.12 Experimental set-up for investigating moments.

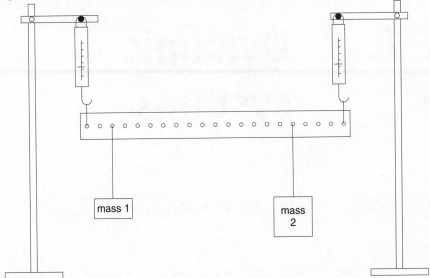

Conclusion

Do your findings confirm the following?

sum of the upward forces = sum of the downward forces

sum of the clockwise moments = sum of the anticlockwise moments

⊶ **Key skills in application of number can be demonstrated in this activity and recorded as portfolio evidence.**

7 Dynamic systems

7.1 Speed, velocity and acceleration

Speed

We indicate how quickly an object is moving by referring to its speed. Speed is the rate at which an object moves from one position to another and is independent of direction.

The formula connecting speed, distance and time is

speed = distance travelled ÷ time taken

If the speed of a body varies over a distance, then the speed will be an average speed.

Units of speed are metres per second (m/s) and kilometres per hour (km/h).

If a car travels 160 km in 120 minutes, then its average speed for the journey can be determined by using the formula

average speed = distance travelled ÷ time taken

Here

distance travelled in metres = 160 × 1000 = 160 000

time taken in seconds = 120 × 60 = 7200

∴ average speed = 160 000 ÷ 7200 = 22.22 m/s.

NOTE ON UNIT CONVERSION

To convert metres per second (m/s) to kilometres per hour (km/h) the following method is used.

- Divide by 1000 to change m to km.
- Multiply by 3600 to change seconds to hours.

Dividing by 1000 and multiplying by 3600 is the same as multiplying by 3.6. Therefore to convert m/s to km/h you need to multiply by 3.6, as in the following example.

22.22 m/s = 22.22 × 3.6 = 79.99 km/h

To convert km/h to m/s the procedure is reversed and can be simplified to dividing by 3.6.

79.99 km/h = 79.99 ÷ 3.6 = 22.22 m/s

Speed has magnitude only and is a scalar quantity. Velocity is the rate of change of distance in a given direction; it is therefore a vector quantity since it has both magnitude and direction.

The units of velocity are the same as those for speed: m/s or km/h.

Uniform velocity

A body moving in a straight line and covering equal distances in equal time is said to be moving with uniform velocity. If a graph is plotted of distance against time for a body moving with uniform velocity, a straight line passing through the origin will be produced. Figure 7.1 shows a distance–time graph for a body moving with uniform velocity.

Fig. 7.1   Distance–time graph for a body moving with uniform velocity.

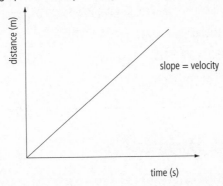

The slope or gradient of the distance–time graph gives the value of the velocity, i.e.

velocity = $\dfrac{\text{change in distance}}{\text{time taken}}$

Acceleration and deceleration

When a body increases its velocity we say it is accelerating. Acceleration is the rate of change of velocity.

Acceleration = change in velocity ÷ time taken

i.e. acceleration = $\dfrac{\text{change in velocity}}{\text{time taken}}$

The units of acceleration are m/s^2 (metres per second, per second).

If a body decreases its velocity it is decelerating or retarding. Using the above formula will produce a negative value. This is acceptable since a negative acceleration is the same as a deceleration.

The velocity–time graphs in Fig. 7.2(a) and (b) show uniform acceleration and uniform deceleration.

Fig. 7.2   Velocity–time graphs showing (a) uniform acceleration and (b) uniform deceleration.

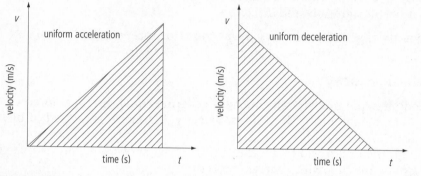

EXAMPLE 7.1

A car increases its velocity from 30 km/h to 50 km/h in 5 seconds. Determine the acceleration.

Acceleration = change in velocity ÷ time taken

Change in velocity = 50 − 30 = 20 km/h = 5.56 m/s   (÷ by 3.6 to convert km/h to m/s)

∴ acceleration = $\dfrac{5.56}{5}$ = 1.112 m/s^2

7.2   Equations of motion

The following equations can be used to solve problems involving linear velocity and linear acceleration:

(i) $v = u + at$

(ii) $v^2 = u^2 + 2as$

(iii) $s = ut + \frac{1}{2}at^2$

(iv) $s = \frac{1}{2}(u + v)t$

The symbols have the following meanings:

u represents initial velocity in m/s

v represents final velocity in m/s

a represents acceleration in m/s^2

t represents time in seconds

s represents distance travelled in metres

The following examples make use of the equations of motion.

EXAMPLE 7.2

A car increases its velocity uniformly from 3m/s to 15m/s in 20 seconds. Determine:

(a) the acceleration;

(b) the distance travelled.

(a) The acceleration can be found by using equation (i) above.

First rearrange $v = u + at$ to make a the subject:

$$a = \frac{(v - u)}{t}$$

Then substituting 3 m/s and 15 m/s for u and v, and 20 seconds for t

$$\text{acceleration} = (15 - 3) \div 20$$
$$= 12 \div 20$$
$$\text{acceleration} = 0.6 \text{ m/s}^2$$

(b) The distance travelled can be found using equation (iii) above.

$$s = ut + \tfrac{1}{2}at^2$$

Substituting in the equation

$$\text{distance travelled} = (3 \times 20) + 0.5 \times 0.6 \times 400$$
$$= 60 + 120$$
$$\text{distance travelled} = 180 \text{ m}$$

EXAMPLE 7.3

A truck travelling at 90 km/h brakes for 15 seconds, reducing its speed to 54 km/h. Determine the distance travelled in the braking period and the retardation value.

$$90 \text{ km/h} = 90 \div 3.6 = 25 \text{ m/s}$$

$$54 \text{ km/h} = 54 \div 3.6 = 15 \text{ m/s}$$

Distance travelled can be found using the equation

$$s = \frac{(u + v)}{2} \times t$$

Substituting the values 25 m/s and 15 m/s for u and v,

$$\text{distance travelled} = \frac{(25 + 15) \times 15}{2}$$

$$= \frac{40 \times 15}{2}$$

$$= 20 \times 15$$

$$\text{distance travelled} = 300 \text{ m.}$$

Retardation can be found using the equation

$$v^2 = u^2 + 2as$$

Rearranging the equation to make a the subject,

$$a = \frac{(v^2 - u^2)}{2s}$$

Then

$$\text{retardation} = \frac{(15^2 - 25^2)}{2 \times 300}$$

$$= \frac{(225 - 625)}{600}$$

$$\text{Retardation} = -0.67 \text{ m/s}^2.$$

The minus value indicates that the velocity is decreasing due to the retardation.

7.3 Velocity–time diagrams

A velocity–time diagram is used to show the relationship between velocity and time for a moving body and can be used to solve problems involving linear motion.

Examples of velocity–time diagrams are given in Fig. 7.3.

Fig. 7.3 Examples of velocity–time diagrams.

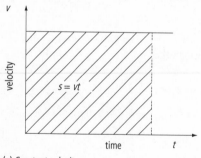

(a) Constant velocity

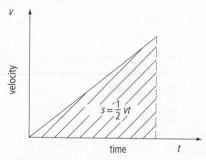

(b) Uniform acceleration from rest

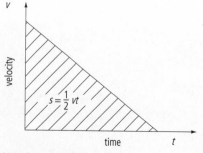

(c) Uniform deceleration to rest

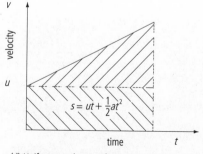

(d) Uniform acceleration from u to v

For the above diagrams the following should be remembered:

- The area under each diagram represents the distances travelled.
- The slope (or gradient) of a velocity–time diagram represents the acceleration.

Velocity–time diagrams can also be used to solve problems involving linear
motion. They can simplify a problem by presenting the data in a way that is
clear and easy to understand. The following examples show the solution to
linear motion problems using both methods.

EXAMPLE 7.4

A vehicle accelerates uniformly from rest to 54 km/h in 8 seconds. Determine the rate of acceleration and the distance travelled while accelerating.

We can convert km/h to m/s by dividing by 3.6; thus 54 km/h = 15 m/s. The velocity–time diagram
is shown in Fig. 7.4.

Fig. 7.4 Finding rate of acceleration and distance travelled from a velocity–time diagram.

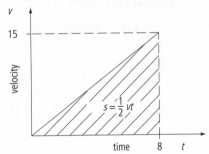

$$\text{acceleration} = \text{slope or gradient} = 15 \div 8 = 1.875 \text{ m/s}^2$$

$$\text{distance travelled} = \text{area under graph} = \text{area of triangle} = \tfrac{1}{2} \text{ base} \times \text{height}$$

$$\text{distance travelled} = 4 \times 15 = 60 \text{ m}$$

Using the equations of motion to solve this problem,

$$s = \tfrac{1}{2}(u + v)t$$

$$s = \tfrac{1}{2}(0 + 15) \times 8$$

$$\therefore s = 60 \text{ m}$$

EXAMPLE 7.5

A car is travelling at 72 km/h and brakes for 20 seconds, reducing its speed to 36 km/h. Determine
the rate of deceleration (retardation) and the distance travelled during braking (note that 72 km/h
= 20 m/s and 36 km/h = 10 m/s).

The velocity–time diagram for this situation is shown in Fig. 7.5.

Fig. 7.5 Finding rate of deceleration and distance travelled while braking from a velocity–time diagram.

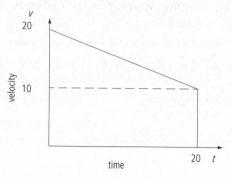

The rate of deceleration = slope of the velocity–time diagram.

$$\text{slope} = \frac{10 - 20}{20} = 0.2$$

∴ deceleration = 0.2 m/s^2

The distance travelled = area under the velocity–time diagram.

distance travelled = area of rectangle + area of triangle

$$= (20 \times 10) + (\tfrac{1}{2} \times 20 \times 10)$$

$$= 200 + 100$$

∴ distance travelled = 300 m

Using the equations of motion to solve this problem, we find a from the equation $v = u + at$.

$$a = \frac{v - u}{t}$$

$$a = \frac{(10 - 20)}{20}$$

∴ acceleration = -0.2 m/s^2, i.e. a deceleration

To find the distance travelled, use $s = \tfrac{1}{2}(u + v)t$.

$$s = \tfrac{1}{2}(20 + 10) \times 20$$

$$= \tfrac{1}{2} \times 30 \times 20$$

∴ distance travelled = 300 m

7.4 Experiments to investigate velocity and acceleration

Two common methods used to investigate motion systems are the ticker tape timer and Fletcher's trolley.

Ticker tape timer

The ticker tape timer is a device which prints a series of dots on a strip of paper pulled through a timing device. The timer can print a dot every 0.02 second (1/50 second). The faster the trolley pulls the tape, the wider the spaces between the dots become. The end result is a tape with a series of dots which can be used to show the acceleration of the trolley.

Figure 7.6 shows a moving trolley set up with ticker tape timer and a typical tape print.

Fig. 7.6   Trolley and ticker tape timer with typical tape print.

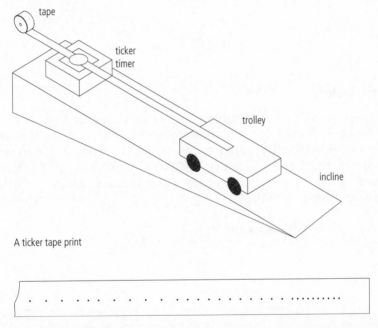

To show that the distance travelled by the trolley varies with time, the tape is cut into sections, each section having an equal number of dots, for example, 5. A dot is printed every 0.02 seconds, so the time period for 5 dots will be $5 \times 0.02 = 0.1$ seconds. Each section will thus represent the distance travelled by the trolley in 0.1 seconds and will therefore show the speed of the trolley.

The tape sections can be made into a tape chart as shown in Fig. 7.7 and a line can be plotted through the dots at the top of each section. This line will represent the change in speed per unit time, i.e. the acceleration of the trolley. The line has a constant gradient and hence shows that the acceleration of the trolley is uniform.

To complete the exercise the acceleration of the trolley can be found by calculating the slope of the graph.

Acceleration = slope = speed ÷ time

Fig. 7.7 Completed tape chart showing the gradient of the graph.

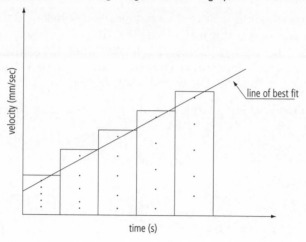

Fletcher's trolley

The second method, Fletcher's trolley, is similar to the first but the timer is replaced by a pen fixed to the end of a spring arm and the track is lowered to a slight gradient before the trolley is pushed with constant rate to simulate the velocity of the trolley. The pen oscillates at the rate of 5 times every second and as the trolley travels a line is drawn in a series of continuous waveforms, as shown in Fig.7.8.

Fig. 7.8 Trace from Fletcher's trolley experiment.

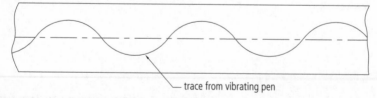

The distance between the peaks of each wave form can be recorded for each 0.2 second time period and a graph plotted for distance per unit time, in order to determine the velocity of the trolley.

The exercise can be repeated by attaching a weight to the trolley via a pulley. The weight is released and accelerates downwards, which accelerates the trolley. A graph produced from the results will show the acceleration of the trolley.

EXAMPLE 7.6

An experiment on a moving trolley produced the following results:

Distance (mm)	0	40	78	122	158	202	245
Time (s)	0	1	2	3	4	5	6

Plot the graph of distance against time and determine the speed of the trolley.

The points have been plotted on the graph in Fig. 7.9 and a line of best fit drawn through the points.

Fig. 7.9 Distance–time graph for a trolley experiment.

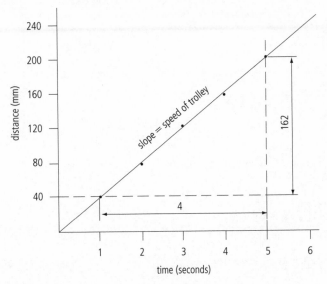

The slope of the distance–time graph represents the speed of the trolley.

$$\text{Speed} = \frac{\text{distance}}{\text{time}}$$

$$= \frac{162}{4}$$

$$= 40.05$$

The speed of the trolley is 40.05 mm/s.

EXAMPLE 7.7

A trolley moving in a straight line produced the following results:

Velocity (m/s)	0	8	16	24	24	24	24
Time (s)	0	1	2	3	4	5	6

Plot the velocity–time graph for the trolley and determine:

  (a) the acceleration of the trolley

  (b) the distance travelled during acceleration

  (c) the distance travelled in 6 seconds

  (d) the average velocity of the trolley.

Figure 7.10 shows the results plotted as a velocity–time graph.

Fig. 7.10 Velocity–time graph for a trolley.

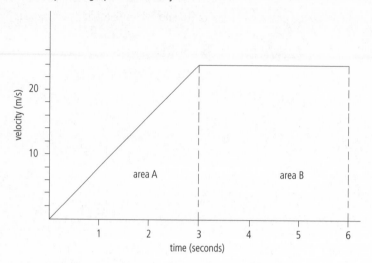

(a) Acceleration = $\dfrac{\text{change in velocity}}{\text{time}}$

$\qquad\qquad\quad = \dfrac{24 - 0}{3} = 8$

∴ acceleration = 8 m/s².

(b) Distance travelled during acceleration = area of triangle A

$\qquad\qquad\qquad\qquad\qquad\qquad\quad = \tfrac{1}{2}\ \text{base} \times \text{height}$

$\qquad\qquad\qquad\qquad\qquad\qquad\quad = 0.5 \times 3 \times 24$

$\qquad\qquad\qquad\qquad\qquad\qquad\quad = 36$

∴ distance travelled during acceleration = 36 m

(c) Distance travelled in 6 seconds = area A + area B

$\qquad\qquad\qquad\qquad\qquad\qquad\ = 36 + (3 \times 24)$

$\qquad\qquad\qquad\qquad\qquad\qquad\ = 36 + 72$

$\qquad\qquad\qquad\qquad\qquad\qquad\ = 108$

∴ distance travelled in 6 seconds = 108 m.

(d) Average velocity = $\dfrac{\text{total distance travelled}}{\text{time taken}}$

$\qquad\qquad\qquad\ = \dfrac{108}{6} = 18$

∴ average velocity = 18 m/s.

EXAMPLE 7.8

A motor cycle accelerates uniformly from rest for 6 seconds until its velocity is 24 m/s. This velocity is maintained for 30 seconds and then the brakes are applied, bringing the machine to rest in 4 seconds.

Sketch the velocity–time graph and determine:

    (a) the acceleration of the motorbike

    (b) the total distance travelled

    (c) the deceleration of the motorbike

    (d) the average velocity.

Figure 7.11 shows the velocity–time graph.

Fig. 7.11   Velocity–time graph for the motorcycle in Example 7.8.

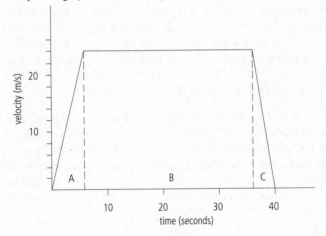

(a) The acceleration of the motorbike $= \dfrac{\text{change in velocity}}{\text{time}}$

$$= \frac{24 - 0}{6} = 4$$

$\therefore$ acceleration $= 4$ m/s^2.

(b) Total distance travelled $=$ area A $+$ area B $+$ area C

$$= (0.5 \times 6 \times 24) + (30 \times 24) + (0.5 \times 4 \times 24)$$

$$= 72 + 720 + 48$$

$$= 840$$

$\therefore$ total distance travelled $= 840$ m.

(c) Deceleration of the motorbike $= \dfrac{\text{change in velocity}}{\text{time}}$

$$= \frac{0 - 24}{4} = -6$$

The − sign indicates that the motorbike is decelerating or slowing down, hence the deceleration of the motorbike is 6 m/s^2.

(d) Average velocity = $\dfrac{\text{total distance travelled}}{\text{time}}$

$$= \frac{840}{40} = 21$$

∴ average velocity = 21 m/s

7.5 Friction, work and power

Friction

When a body moves or a force is applied to a body to start movement, there are forces which oppose the motion or the tendency of the body to move. These forces are known as friction forces and oppose the relative motion between two surfaces in contact. The resistance to motion (friction) varies depending on the roughness of the surfaces in contact, the materials of the surfaces and whether the surface is dry or lubricated.

In engineering, friction can be an essential factor in a machine where it is used for clamping devices, belt drives, clutch plates and brakes. In other cases friction needs reducing as much as possible because of the heat that is generated when surfaces rub together, for example in gear boxes, bearings and slides on machine tools. Friction forces always oppose the motion of a body, and therefore act in the opposite direction to the applied force, and to the direction of motion.

The magnitude of the friction force acting between two bodies in contact depends on:

- the normal reaction N between the two surfaces
- the nature and materials of the two surfaces
- the speed at which the surfaces move over one another

The area of the surfaces in contact does not affect the friction force, providing it is not so small that it would cause penetration of the surfaces.

STATIC FRICTION

Figure 7.12 shows an object resting on a surface. The weight of the object exerts a downward force of mg and the surface applies an equal and opposite reaction to the force. This is known as the normal reaction N.

When the horizontal force F is applied, it is opposed by friction force F_F. As the horizontal force increases, the friction force will also increase until the object is on the point of moving. This type of friction is known as static friction, as it is the friction resisting motion in a body at rest.

Fig. 7.12   Static friction.

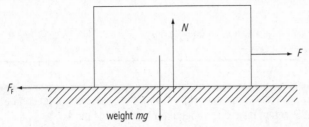

weight *mg*

For dry clean surfaces a simple relationship exists where the friction force is directly proportional to the normal reaction between the two surfaces, i.e.

$F \propto N$ (the symbol $\propto$ means proportional to)

or F = a constant value $\times N$

The constant value is known as the coefficient of friction and is identified by the symbol μ (mu).

$\therefore F = \mu N$

or $\mu = \dfrac{F}{N}$

The coefficient of friction has no units as it is a ratio of similar units. The value of the coefficient will depend on the nature of the materials and the surface finish of the surfaces in contact.

Common average values for μ, for dry unlubricated surfaces, are shown in Table 7.1.

Table 7.1   Values of the coefficient of friction for dry, unlubricated surfaces

Material	Coefficient of friction μ
Metal on metal	0.2
Rubber on concrete	0.8
Brake lining on cast iron	0.4
Leather on steel	0.45
Rubber on steel	0.6

EXAMPLE 7.9

A body of mass 200 kg rests on a horizontal surface. If the force required to move the body initially is 400 N, determine the coefficient of friction for the surfaces in contact.

$N = mg = 200 \times 9.81 = 1962\ N$

Coefficient of friction $\mu = \dfrac{F}{N}$

$$\therefore \mu = \frac{400}{1962} = 0.2$$

The coefficient of friction for the two surfaces is 0.2.

EXAMPLE 7.10

A block of mass 100 kg rests on a table (Fig. 7.13). If the coefficient of friction between the table and the block is 0.25, determine the force required to move the block.

Fig. 7.13 Static friction between a block and a table.

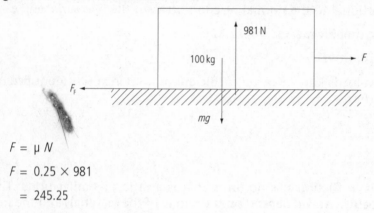

$F = \mu N$

$F = 0.25 \times 981$

$= 245.25$

The force required to move the block is 245.25 N.

DYNAMIC FRICTION

When a body overcomes the initial force resisting motion (the static friction) it will move more easily and less effort is required. The force resisting motion for the moving body is called dynamic friction and the coefficients of friction are slightly higher than the values for static friction.

EXAMPLE 7.11

A block of mass 50 kg is pulled along a horizontal table by a force of 200 N (Fig. 7.14). Determine the coefficient of dynamic friction between the two surfaces.

Fig. 7.14 Dynamic friction between a block and a table.

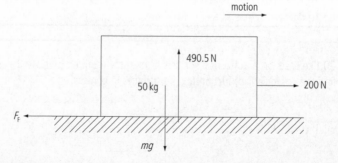

Coefficient of friction $\mu = \dfrac{F}{N}$

$$= \dfrac{200}{490.5} = 0.41$$

$\therefore$ The coefficient of dynamic friction is 0.41.

EXAMPLE 7.12

A mass of 100 kg rests on a horizontal table. If the coefficients of static and dynamic friction between the two surfaces in contact are 0.3 and 0.25, respectively, determine:

(a) the force required to cause the mass to move from rest

(b) the force required to keep the mass moving at a constant speed.

The force required to move the body from rest is given by $F = \mu N$, where μ is the coefficient of static friction. In this case $N = mg = 100 \times 9.81 = 981$ N

$\therefore F = 0.3 \times 981$

$= 294.3$ N

The force required to keep the mass moving at constant speed is given by $F = \mu N$, where μ is the coefficient of dynamic friction.

$F = 0.25 \times 981$

$= 245.25$ N

The mass requires a force of (a) 294.3 N to move it initially and a force of (b) 245.25 N to keep it moving at constant speed.

Work

Work is defined as the force required to overcome resistance to motion multiplied by the distance moved in the direction of the force, i.e.

work = force $\times$ distance

The unit of work is the joule, symbol J. 1 joule is the work done when a force of 1 newton moves its point of application through a distance of 1 metre. Hence 1 J = 1 Nm.

If a force is applied and there is no motion, then there is no work done, even if the magnitude of the force is very large.

EXAMPLE 7.13

A force of 200 N is applied to a trolley, moving it a distance of 4 m in the direction of the force. Determine the work done on the trolley.

Work = force $\times$ distance

where force = 200 N and distance = 4 m,

$$\therefore \text{work} = 200 \times 4$$

$$= 800 \text{ J}$$

The work done on the trolley is 800 J (or N m since 1 J = 1 N m).

EXAMPLE 7.14

The work done on a body is 1200 J, causing the body to move 20 m in the direction of an applied force. Determine the applied force.

Work = force × distance

$$\therefore \text{force} = \frac{\text{work}}{\text{distance}}$$

$$\therefore \text{force} = \frac{1200}{20} = 60$$

The force applied to the body = 60 N.

WORK DIAGRAMS

If a graph is drawn of force against distance, the area under the graph will represent the work done by the force. The force–distance graph is called a **work diagram**.

EXAMPLE 7.15

A force of 20 N moves an object a distance of 10 m. Construct a work diagram and determine the work done.

Figure 7.15 shows the work diagram.

Fig. 7.15 Work diagram.

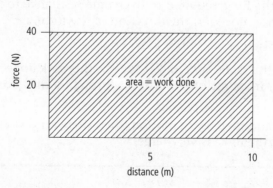

Work done = area under graph

$$= 40 \times 10$$

$$= 400$$

The work done by the force = 400 J

EXAMPLE 7.16

A pulley is used to lift a crate through a distance of 5 m with an applied effort of 20 N (Fig. 7.16). Determine the work done on the crate.

Fig. 7.16 Pulley used to lift a crate.

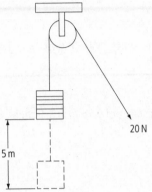

Figure 7.17 shows the work diagram.

Fig. 7.17 Work diagram for lifting the crate.

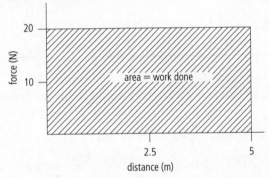

Work done in raising crate = area under graph

$$= 20 \times 5$$

$$= 100$$

The work done in raising the crate is 100 J

Energy

Energy is the ability to do work. When energy is supplied to a system it allows work to be done. In an ideal system all the energy will be converted to work. However, friction and other forms of energy will be lost and there will always be less energy to do useful work than the original amount available.

Power

Power is the rate of doing work.

$$\text{Power} = \frac{\text{work done}}{\text{time taken}}$$

$$P = \frac{w}{t}$$

The unit of power is the watt (W). 1 watt = 1 Joule per second (J/s).

EXAMPLE 7.17

A shaping machine moves a distance of 0.75 m in 3 seconds. If the force on the tool is 1.5 kN, determine the power required.

$$\text{power} = \frac{\text{work done}}{\text{time taken}}$$

$$= \frac{\text{force} \times \text{distance}}{\text{time}}$$

$$= \frac{1500 \times 0.75}{3} = 375$$

The power required is 375 W.

EXAMPLE 7.18

Calculate the work done and the power required to raise a 250 kg mass through a height of 60 m in 50 seconds.

The work done can be found from

$$\text{Work done} = \text{force} \times \text{distance}$$

where force $= mg$

$$= 250 \times 9.81$$

$$= 2452.5 \text{ N}$$

$$\therefore \text{work done} = 2452.5 \times 60$$

$$= 147\ 150 \text{ N}$$

Having found out how much work is done, the power can be found from

$$\text{power} = \frac{\text{work done}}{\text{time taken}}$$

$$= \frac{147\ 150}{50} = 2943$$

$\therefore$ power = 2943 W or 2.943 kW.

EXAMPLE 7.19

Figure 7.18 shows the force required to move a load across a horizontal floor. Determine the work done and the power required to move the load in 20 seconds.

Fig. 7.18   Force required to move a load across a horizontal floor.

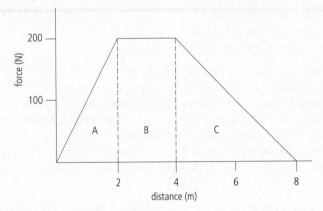

From the work diagram,

work done = area under graph

= area A + area B + area C

$$= \frac{(2 \times 200)}{2} + (2 \times 200) + \frac{(4 \times 200)}{2}$$

= 200 + 400 + 400

= 1000 J

The work done is 1000 J or 1 kJ.

The power is given by

$$power = \frac{work\ done}{time\ taken}$$

$$= \frac{1000}{20} = 50$$

The power required to move the load in 20 s is 50 W.

Chapter review

Speed   we indicate how quickly an object is moving by referring to its speed

Velocity   how quickly an object moves in a specified direction

Acceleration   the rate of change of velocity

Retardation   if the velocity of a body decreases, the body is decelerating or retarding

Velocity–time diagrams   used to show the relationship between velocity and time for a moving body and can be used to solve problems involving linear motion

Equations of motion   formulae used to calculate values such as distance, velocity, acceleration and deceleration. In these equations:

   u represents initial velocity in m/s
   v represents final velocity in m/s
   a represents acceleration in m/s^2
   t represents time in seconds
   s represents distance travelled in metres

Friction   the resistance to motion of a body

Static friction   the force resisting motion in a body at rest

Dynamic friction   the force resisting motion in a moving body

Work   work is done when a force applied to a body moves the body in the direction of the force. The unit of work is the joule, where $1\text{ J} = 1\text{ N m}$

Energy   the ability to do work. Energy has the same units as work, joules (J)

Power   the rate of doing work. Power is measured in watts ($1\text{ W} = 1\text{ J/s}$)

Exercise 7.1

Note: remember to convert distances to metres and speeds or velocities to metres per second where necessary.

1 A forklift truck moves 600 m in 2 minutes. Calculate the speed of the truck.

2 A car reaches a speed of 70 km/h from rest in 6 seconds. Determine the rate of acceleration.

3 A car under test conditions produced the following speeds and times.

Time (s)	0	5	10	15	20	25	30
Speed (m/s)	0	10	20	30	40	50	60

Plot a speed–time graph of the results and determine the acceleration and the overall distance travelled in the test.

4 A train travels 500 km at an average speed of 80 km/h. Determine the time taken to travel this distance.

5 A vehicle accelerates uniformly from 20 km/h to 80 km/h while travelling a distance of 360 m. Sketch a velocity–time diagram and determine the acceleration and the time taken.

6 A vehicle is brought to rest from 70 km/h in 12 seconds. Determine the retardation and the distance travelled in coming to rest.

7 Define velocity and acceleration and state the SI units for each.

8 The speed of a car in a busy city centre is shown in the table.

Time (s)	0	10	15	20	40	50	70
Speed (m/s)	0	3	4.4	6	6	4	0

Plot a speed–time graph for the car and determine:

(a) the acceleration of the car

(b) the deceleration of the car

(c) the distance travelled during acceleration

(d) the total distance travelled.

9 A hammer falls from a scaffold walk and takes 4 seconds to hit the ground. Assuming that the hammer accelerates at a rate of 9.81 m/s^2, determine the height of the scaffold.

10 An object on a conveyor moves as follows:

- from rest to 10 m/s in 30 seconds
- this speed is maintained for 3 minutes then
- the conveyor comes to rest in 18 seconds.

Determine:

(a) the rate of acceleration

(b) the rate of deceleration

(c) the total distance the object travels.

11 A rope is used to pull a casting across a horizontal surface. If the mass of the casting is 500 kg, determine the force required to initially move it, if the coefficient of friction between the surfaces in contact is 0.2.

12 A force of 150 N is used to push a tailstock along a lathe bed. If the coefficient of friction between the tailstock and the lathe bed is 0.25, determine the mass of the tailstock.

13 A force of 250 N is used to lift a box on to a table. If the table is 1.2 m high and the box was initially on the floor, determine the work done in lifting the box.

14 Figure 7.19 shows a distance graph for a load. Determine the work done and the power required to move the load in 12 seconds.

15 A load of mass 1000 kg is raised through a height of 12 m in 20 seconds. Determine the power.

16 A lift carries a 500 kg load through a vertical distance of 60 m in 30 seconds. Calculate the power.

Fig. 7.19 Force–distance graph for a load.

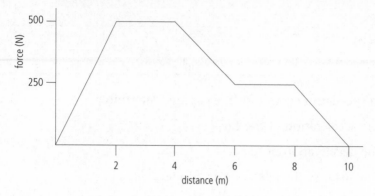

Activity sheet

Dynamic systems

Conduct an experiment using a moving trolley to investigate velocity and acceleration for a body in motion system.

You will need the following equipment:

- a moving trolley with ticker timer or Fletcher's trolley
- a weight carrier and some suitable weights to accelerate the trolley
- a rule, scissors, graph paper, pencil and a calculator might be useful.

Draw a labelled diagram of the equipment set up and describe in detail the method used in this experiment.

Produce graphs to show uniform velocity and uniform acceleration for the trolley and determine the value of the gradients of the graphs.

Use the equations of motion to verify the results for uniform velocity and uniform acceleration.

Compare the area under the constant velocity–time graph (i.e. distance travelled) with the actual distance travelled by the trolley on the plane. Comment on the values and explain any differences (if any) in the values.

Further tests

If you wish you could vary the angle of the trolley plane and repeat the experiment, or add weights to the trolley to increase its mass and repeat, noting the results, drawing graphs and comparing values.

○→ Key skills in application of number and information technology can be demonstrated in this activity and recorded in evidence portfolios.

chapter

8

Expansion of materials

8.1 Physical changes

When a solid object is heated, various changes may take place. The size, appearance or hardness may change, but the material is still the same. These changes are known as **physical changes**.

Experiments are quite easy to perform to investigate the physical change of a material when the temperature changes and may be carried out in a laboratory or engineering workshop.

In engineering it is common to investigate linear expansion or contraction, which is the effect of temperature on the linear dimensions of a solid material, and the volumetric expansion or contraction, which is the effect of temperature on a the volume of a liquid. In all cases the expansion or contraction of a body is related to its temperature.

Linear effects

A railway line extends when hot and contracts (becomes smaller) when cold. Expansion joints are fitted to railway tracks to prevent them warping due to the heat or cracking due to tension when contracting.

Volumetric effects

A liquid will expand when hot, such as the alcohol in a thermometer, and will contract as the temperature decreases.

8.2 Heat

Heat is a form of energy which is obtained when other forms of energy are converted. For example chemical energy is released when a match is struck. The match head ignites and the substance reacts with oxygen from the air to release chemical energy. The chemical energy is converted into heat and light energy, which are different kinds of energy to the chemical energy. The process whereby energy is converted from one form to another is known as the conversion of energy. Although energy can be converted, or changed, from one form to another, it cannot be created or destroyed. This is stated in the law of conservation of energy.

The SI unit for heat energy is the joule, abbreviated to J.

Figure 8.1 shows energy changing from one form to another.

Fig. 8.1 Conversion of energy.

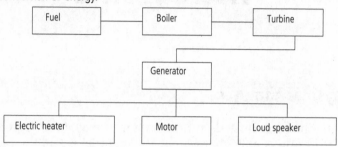

Heat transfer

Heat energy may be transferred by conduction, convection and radiation.

Conduction is the transfer of heat energy from one part of the body to another without the particles of the body moving. If a metal rod is heated at one end, the other end will become hot because of conduction. Conduction is associated with solids such as metals, which are good conductors of heat. Wood, glass and plastics do not conduct heat and are known as insulators.

Convection is heat transfer through a substance as a result of the movement of the molecules within it. When water in a kettle is heated from the bottom, the liquid at the bottom becomes warmer and less dense. The hotter, lighter water rises, and cooler, denser water moves downward to take its place. In a hot water tank, water is usually drawn off from the top. This is because the water rises to the top of the tank when it is heated, so the water at the top is hotter than that at the bottom.

Radiation is the transfer of heat energy from a hot body to a cooler one by electromagnetic waves. Radiant heat is reflected from shiny surfaces, but absorbed by dull, black surfaces. Heat energy from the sun reaches us by radiation. The sun emits electromagnetic waves that travel at high speed through space. When the heat waves reach a body some may be reflected, while some are absorbed, producing a rise in temperature. The heat emitted by an electric fire is reflected from a shiny curved panel behind the heating element and travels through the air to warm the surrounding room.

When heat is either supplied to or taken from a body resulting in a temperature change, it is known as **sensible heat**.

8.3   Temperature

This is the hotness of a body measured on an agreed scale. We are familiar with degrees Celsius (°C) and usually state the temperature of a body in this way. Thermometers record temperature in °C and °F (Fahrenheit) and these units are used in normal daily applications. For example:

- the normal body temperature is accepted as 98.6°F
- water boils at 100°C and freezes at 0°C
- our daily weather forecasters use °C and °F when informing us of the expected maximum and minimum temperatures.

The kelvin temperature scale

Engineers use a different unit to measure temperature for thermodynamic systems. Boilers and turbines are examples of thermodynamic systems and use the kelvin (K) for measuring temperature, which is the SI unit. Zero kelvin (0 K) is known as absolute zero and is the temperature at which all the internal energy has been extracted from a body. Absolute zero occurs 273° below 0°C, i.e. −273°C = 0 K.

Figure 8.2 shows a turbine, which uses the energy in flowing steam to drive a generator. A turbine is a thermodynamic system.

Fig. 8.2   Steam turbine.

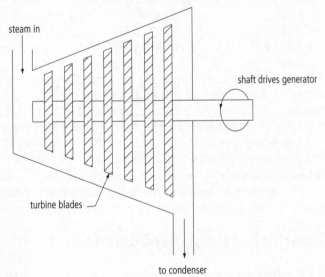

steam in

shaft drives generator

turbine blades

to condenser

Thermometers

An instrument which is used to measure temperature is called a thermometer. One of the most common methods of measuring temperature is to make use of the fact that most substances expand on heating and contract on cooling. The alcohol and mercury-in-glass thermometer, shown in Fig. 8.3, work on this principle.

Fig. 8.3 The thermometer.

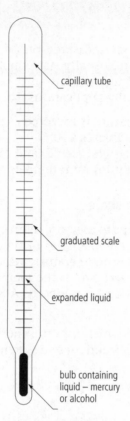

capillary tube

graduated scale

expanded liquid

bulb containing
liquid – mercury
or alcohol

Mercury is used because it remains liquid over a considerable range of temperature, from −39°C to 357°C. The sensing bulb contains a volume of mercury which, on expansion, can move up a capillary tube (a very fine bore tube of uniform cross-section) which enables small changes of volume to be accurately measured. However, mercury freezes at −39°C, which makes its application limited. For instance, temperatures in the Arctic drop well below −39°C and the mercury would freeze. Alcohol freezes at −115°C, so it would be better suited to Arctic conditions. In hot climates a mercury thermometer would be better suited since it remains a liquid up to 357°C. Alcohol boils at about 78°C and therefore would be unsuitable in hot climates.

8.4 Linear expansion of solid materials

Most materials expand as the temperature increases and contract (become smaller) as the temperature decreases. Figure 8.4 shows a bar expanding and contracting with changes in temperature.

The change in length x of the material depends upon the following:

- the original length l
- the change in temperature Δt (final temperature − initial temperature)
- the coefficient of linear expansion of the material α

Fig. 8.4   Expansion and contraction with change in temperature.

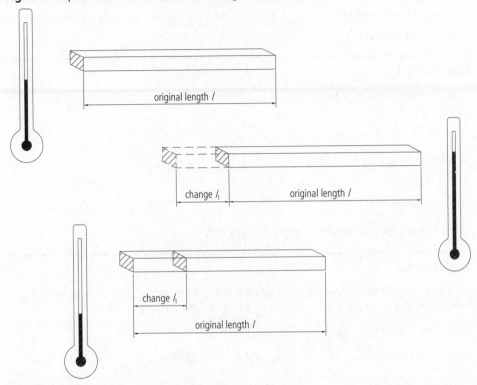

8.5   Coefficient of linear expansion

This is the amount of change in length for every 1°C change in temperature and is different for each material. It is identified by the symbol α (alpha).

Table 8.1   Coefficients of linear expansion for common materials

Material	Coefficient of linear expansion α
Aluminium	$24 \times 10^{-6}/°C$
Brass	$19 \times 10^{-6}/°C$
Bronze	$18 \times 10^{-6}/°C$
Copper	$16 \times 10^{-6}/°C$
Nylon	$100 \times 10^{-6}/°C$
Rubber	$80 \times 10^{-6}/°C$
Steel	$12 \times 10^{-6}/°C$

The change in length of a material can be determined using the following formula:

change in length = original length × coefficient of linear expansion × temperature change

i.e. $x = l\alpha\Delta t$

EXAMPLE 8.1

A copper tube at 20°C is 3 m in length. Determine the increase in length if the temperature of the pipe is raised to 28°C.

Change in length = $l\alpha\Delta t$

where $l = 3$ m, $\alpha = 16 \times 10^{-6}/°C$ and $\Delta t = 28 - 20 = 8°C$,

$\therefore x = 3000 \times 16 \times 10^{-6} \times 8$

$= 0.384$

i.e. the copper pipe will increase in length by 0.384 mm.

EXAMPLE 8.2

A steel rail is 30 m long when the temperature is 60°C. If the temperature drops to −10°C on a frosty night, determine the decrease in length. Take α for steel from Table 8.1.

Change in length = $l\alpha\Delta t$

where $l = 30$ m, $\alpha = 12 \times 10^{-6}/°C$ and $\Delta t = 10 - 60 = -70°C$,

$\therefore x = 30 \times 12 \times 10^{-6} \times -70$

$= -0.0252$ m

i.e. the rail decreases in length by 25.2 mm.

8.6 Engineering applications for thermal expansion of materials

Bearings can be heated until they expand in diameter. This allows them to be easily fitted to shafts. When the bearing cools the material contracts and is fixed to the shaft (Fig. 8.5).

Fig. 8.5 Fitting a bearing to a shaft.

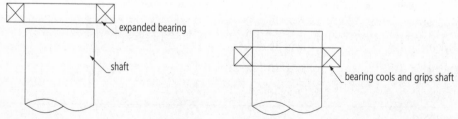

Shrink fitting of internal components such as cylinder linings and bearings can also be achieved. In these applications the bearing is cooled to a very low temperature using liquid nitrogen which contracts the material. The lining or bearing can then be fitted easily into the housing. As the material returns to the room temperature it expands to its original length and compresses against the housing wall (Fig. 8.6). This is known as a shrink fit.

Fig. 8.6 Shrink fitting internal components.

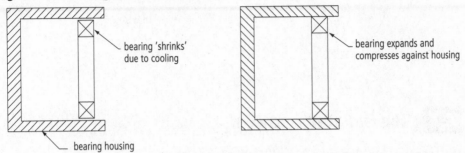

bearing 'shrinks' due to cooling

bearing expands and compresses against housing

bearing housing

The effect of thermal movement must always be considered in engineering and industrial design. In an engine, moving parts, such as the pistons in the cylinder block and the associated timing and valve components, must be designed to allow for expansion under operating conditions. Pendulums and balance wheels of watches depend on accurate dimensions and therefore expansion and contraction effects must be minimized to avoid time errors. This is normally accomplished by either using specially developed alloy materials which have a very low expansion coefficient, or by compensation techniques where the expansion of one component counteracts the expansion of another.

In bridge and building construction, expansion joints must be provided. Bridge structures and roads are fitted with interlocking joints to allow for movement due to temperature changes throughout the year. Railway tracks are also fitted with expansion joints. The steel rails must be laid with gaps between them, otherwise a temperature rise would cause severe buckling owing to the compression forces involved when one rail expands against another. Bridges have rolling supports to allow for temperature changes as the expansion can be large for such a long structure.

In accurate measurement using steel rules, micrometers, etc., the temperature of measurement must be noted and corrections made for expansion or contraction in these instruments, if measurements are taken at a temperature different from the temperature of their calibration. Measuring instruments are normally calibrated at 20°C. Care should be taken when using precision measuring instruments such as the micrometer, as temperature rises will affect the true reading. In the case of the micrometer it should always be held with the palm of the hand supporting the frame and never held by the spindle, as this may cause a increase in length. Finally, a micrometer should not be handled for longer than is necessary to take the reading.

Fig. 8.7 The micrometer.

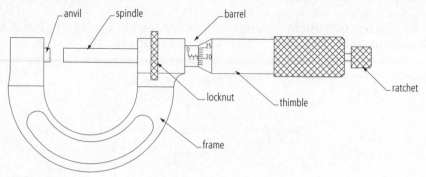

8.7 The bimetallic strip

Different materials expand by different amounts. A brass bar will expand by a greater amount than an iron bar of the same length for the same temperature rise. This is because brass expands more than iron.

A bimetallic strip (Fig. 8.8) is made from a strip of brass fixed to a strip of iron. When the strip is heated the brass expands more than the iron and the strip bends. As the strip cools, the brass contracts more than iron and it bends the other way. Thermostats use bimetallic strips to control central heating systems and can also be used to detect temperature changes in buildings to operate fire alarm systems.

Fig. 8.8 The bimetallic strip.

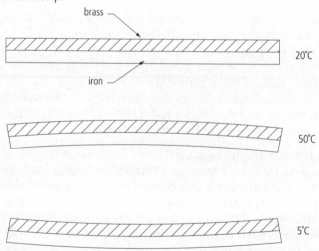

Room thermostat

As the temperature in the room increases the bimetallic strip expands, bending away from the contact. The circuit becomes open and the heater unit switches off. As the room cools the strip returns to its normal length and makes contact, closing the circuit and switching the heater on again.

Fig. 8.9 The room thermostat.

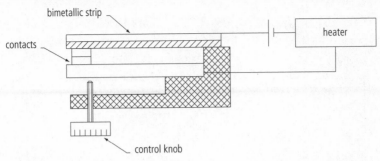

Alarm

Fig. 8.10 The alarm.

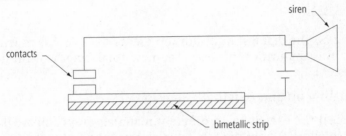

In the alarm circuit the bimetallic strip bends towards the contact when it expands due to temperature increase. This closes the circuit and operates the alarm.

8.8 Practical investigation into the physical changes of a material

You will need:

- a length of rod such as steel, copper, brass or aluminium
- a flame torch or Bunsen burner
- a workshop vice
- a dial test indicator (D.T.I.) and a steel rule

Set the equipment up as shown in Fig. 8.11. Measure the length of the rod and record for reference.

Position the dial test indicator against the end of the bar and reset to zero.

Apply the flame from the torch to the rod for about 1 minute.

Observe the reading on the dial test indicator, showing that the rod is increasing in length.

Remove the flame and record the reading on the dial test indicator.

Fig. 8.11 The dial test indicator.

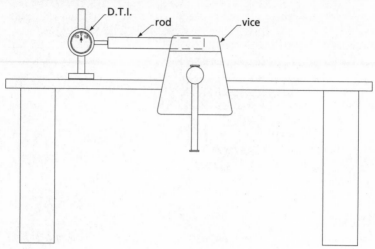

The reading on the dial test indicator represents the increase in length of the rod due to the application of heat. The new total length of the rod can be found thus:

total length = original length + extension

You can repeat the exercise with different materials and compare the changes in length, showing the results in a table or graph of materials and respective changes in length.

8.9 Specific heat capacity

The specific heat capacity of a substance is the amount of heat energy required to raise the temperature of one kilogram of a substance by one degree (kelvin or celsius). The symbol 'c' is used when referring to specific heat capacity in formulae. Specific heat capacity is measured in joules per kilogram per kelvin (J/kg/K).

The specific heat capacity is used to determine the amount of heat energy required to raise the temperature of a substance. Typical values for some common substances are shown in Table 8.2.

The amount of heat energy required to raise the temperature of a substance may be calculated using

$Q = mc\,(t_2 - t_1)$

where Q = heat required

m = mass of substance in kilograms

t_1 = initial temperature

t_2 = final temperature

c = specific heat capacity of the substance.

Table 8.2 Specific heat capacities of common substances

Substance	Specific heat capacity (J/kg/K)
Steel	486
Aluminium	908
Water	4186
Ice	2000
Copper	385
Iron	460
Mercury	139

EXAMPLE 8.3

Determine the heat energy required to raise 2 litres of water initially at 20°C to a final temperature of 56°C. (note that 1 litre of water = 1 kg).

$$\text{Heat energy} = Q = mc(t_2 - t_1)$$
$$= 2 \times 4186 \times (56 - 20)$$
$$= 2 \times 4186 \times 36$$
$$= 301\ 392\ \text{J}$$
$$= 301.4\ \text{kJ}$$

EXAMPLE 8.4

Calculate the energy required to raise the temperature of a 2 kg aluminium ingot initially at 10°C to a temperature of 45°C.

$$\text{Energy} = Q = mc(t_2 - t_1)$$
$$= 2 \times 908 \times (45 - 10)$$
$$= 2 \times 908 \times 35$$
$$= 63\ 560\ \text{J}$$
$$= 63.56\ \text{kJ}$$

When a substance is heated inside a tank, the heat required to raise the temperature of the tank must also be considered.

EXAMPLE 8.5

Calculate the heat energy required to raise the temperature of 100 kg of water in a copper tank of mass 10 kg from 20°C to 56°C.

To solve the problem we can determine the heat required for the two substances individually:

Heat required to raise water temperature $= Q = mc(t_2 - t_1)$

$$= 100 \times 4186 \times (56 - 20)$$

$$= 100 \times 4186 \times 36$$

$$= 15\ 069\ 600\ \text{J}$$

Heat required to raise copper tank temperature $= Q = mc(t_2 - t_1)$

$$= 10 \times 385 \times (56 - 20)$$

$$= 10 \times 385 \times 36$$

$$= 138\ 600\ \text{J}$$

Total heat energy $= 15\ 069\ 600 + 138\ 600$

$$= 15\ 208\ 200\ \text{J}$$

$$= 15.21 \times 10^6\ \text{J}$$

$$= 15.21\ \text{MJ}$$

Total heat energy $= 15.21$ MJ.

8.10 Efficiency

$$\text{efficiency} = \frac{\text{useful output energy}}{\text{input energy}}$$

When energy converts from one form to another, some of the energy may be transferred to the surrounding environment. The ratio of energy output to energy input is known as the efficiency. For example, when an electric kettle heats water, the kettle material – steel, copper or plastic – also gets hot. This heat is conducted to the atmosphere and wasted. The wastage may be reduced by using a material which is a poor conductor of heat for the kettle body, which is difficult to achieve. On a larger scale copper water tanks are insulated to prevent heat loss and insulating jackets are also available. Copper pipes may be lagged to prevent heat loss which can waste energy. In winter unlagged pipes may freeze, resulting in a burst pipe.

EXAMPLE 8.6

An electric kettle is used to raise 2 litres of water from 20°C to 100°C. Determine the energy required to heat the water and the efficiency of the kettle if 900 kJ of electrical energy was used to boil the water.

Heat energy $Q = mc(t_2 - t_1)$

$$= 2 \times 4186 \times (100 - 20)$$

$$= 2 \times 4186 \times 80$$

$$= 669\ 760\ \text{J}$$

Efficiency $=$ $\dfrac{\text{useful output energy}}{\text{input energy}}$,

where useful output energy $=$ 669 760 J and energy supplied $=$ 900 000 J,

$\therefore$ efficiency $=$ $\dfrac{669\ 760}{900\ 000}$

$\qquad\qquad = 0.744$

We can state this value as a percentage by multiplying by 100, i.e.

efficiency $= 0.744 \times 100 = 74.4\%$

Thus the efficiency of the kettle is 74.4%.

8.11 Change of state – latent heat

Matter may exist in three states:

1 solid
2 liquid
3 gas

If we apply heat to a solid its temperature will increase. The temperature will continue to rise until the solid begins to melt and the substance becomes liquid.

Suppose some crushed ice, initially at $-10°C$, is heated in a beaker, as illustrated in Fig. 8.12. The ice is heated very slowly, and is stirred thoroughly

Fig. 8.12 Observation of temperature change as ice is slowly heated.

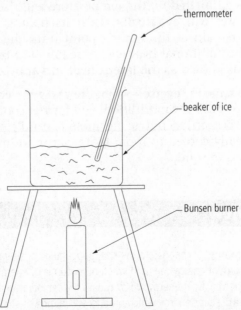

thermometer

beaker of ice

Bunsen burner

before each reading of the thermometer. The temperature on the thermometer is noted at regular intervals as the ice melts. Once melted the water is heated until it boils, with the temperature reading noted at regular time intervals throughout the process.

Figure 8.13 shows the results of such an experiment. At point (a) the ice is solid at −10°C. From (a) to (b) the ice is gradually being heated, so that at (b) the ice is at 0°C. At this point the temperature does not rise, even though the beaker is still being heated. The temperature remains at 0°C but the ice now melts to form water. The process of changing from the solid state at (b) to the liquid state at (c) requires heat energy. This energy is called the **latent heat of fusion** (melting).

Fig. 8.13   Results of heating experiment.

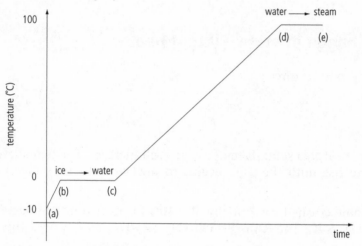

When the ice has all melted (c) the temperature will rise again. If more heat is applied the water will eventually reach its boiling point (d). Here the temperature will remain at the boiling point until the water changes state once more (e). The additional heat energy required to turn water into steam (its vapour form) is known as the **latent heat of vaporization**.

The process is the same in reverse – when the gas is cooled it becomes a liquid, giving out its latent heat of vaporization, and further cooling produces a solid. Experiments can be carried out using molten candle wax or naphthalene (mothballs) gradually cooled to room temperature to investigate the change of state from liquid to solid.

Chapter review

Physical change   changes to the size, appearance or shape of a solid
Heat   a form of energy released when other forms of energy are converted
Sensible heat   heat energy which causes the temperature of a body to change
Latent heat   heat energy which produces a change of state

Temperature the hotness of a body on an agreed scale
Kelvin the SI unit of temperature
Linear expansion the expansion or contraction of a material caused by
temperature change
Coefficient of linear expansion the amount of change in length for every one
degree kelvin
Specific heat capacity the heat energy required to raise the temperature of a
substance by one degree kelvin
Efficiency the ratio of useful energy output to energy input

Exercise 8.1

(Use the coefficients from Table 8.1 when answering these questions.)

1 The steel plate shown in Fig. 8.14 has a hole 12 mm diameter drilled
through the centre. If the temperature of the plate is raised by 20°C, determine the increase in hole size.

Fig. 8.14 Expansion of a steel plate with a hole in it.

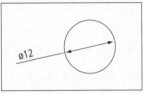

2 A steel rail is 30 m long when its temperature is 12°C. Determine the
increase in a series of 30 rails when the temperature rises to 80°C.

3 A copper tube is 3m long at 20°C. Determine the length of the tube when
the temperature is 40°C.

4 With the aid of diagrams describe what is meant by the term coefficient of
linear expansion.

5 Describe how the application of heat can be used to fit a bearing to a shaft.

6 A steel rod and collar are the same diameter and a push fit is possible when
the temperature of the two components is the same.

Describe what happens if the rod is:

(a) heated to a higher temperature;
(b) cooled to a lower temperature.

7 A copper pipe in a central heating system is 2 m long when the water
temperature is 12°C. Determine the length of the pipe when the boiler
operates and raises the water temperature to 56°C.

8 A steel beam is 20 m long at 20°C. Determine its length at −20°C, 0°C and
40°C.

Activity sheet 🖎📁

Heat and energy

You will need:

- a thermometer

- a measuring jug

- two kettles

- a stopwatch

1 Determine the heat energy required to boil 2 litres of water for the following kettles:

   (a) a modern jug kettle

   (b) a traditional steel kettle

2 Determine the electrical energy used to boil the two kettles using the formula described in Chapter 9.

3 Determine the efficiency of each kettle and state which one is the more economical to use.

4 Produce a pie chart to show the energy provided/used to boil each kettle.

5 Which kettle would you choose if you were buying a kettle? Support your answer with the reasons for your choice.

6 Draw a cross-sectional diagram of a kettle and show the main components. Describe how the cutoff switch operates.

Further investigation

Obtain an old or disused kettle and dismantle it. Investigate the materials it is made from and how its cutoff switch operates.

---

   o— **Key skills in application of number and information technology skills can be demonstrated with this activity and recorded as portfolio evidence.**

chapter

9 Electrical systems

9.1 Introduction to electricity

Electricity is used every day by most people in the world. In the home electricity is used to make domestic tasks such as washing clothes and dishes easier, by allowing us to use washing machines and dishwashers. Food can be prepared and cooked using electrical appliances, homes are lit and warmed using electricity, and all industry uses electricity to manufacture its products. Public transport uses electric motors and lights and communication systems use computers. In the home electricity is supplied on a 230 V alternating current (a.c.) system. In industry and on-site work the electricity supply is often a 110 V direct current (d.c.) system. Electricity can also be supplied as d.c. from a battery. In whatever form, electricity is a product which we all take for granted, and which would be impossible to replace by any other form of energy, while providing the same versatility.

In the study of electrical systems we are concerned with electric current as a fundamental or basic quantity. The unit of current is the ampere (symbol A); since this is a base unit, other units are derived from it.

9.2 Electrical conductors and insulators

An electric current is a flow of electrons through a **conductor**. An example of a conductor of electric current is copper. Copper is classed as a good conductor as there are many free electrons in copper which can carry charge. Glass and rubber have few free electrons to carry a charge and are poor conductors of electric current. These materials are referred to as **insulators** and they are used to cover electric conductors such as copper wire, or for casings of electric plugs.

- Good conductors include: copper, silver and aluminium
- Insulators include: glass, ceramics and PVC

9.3 Basic electric circuits

A battery-powered torch is an example of a basic electric circuit. When a battery (or cell) is connected across the ends of a conductor, the negatively charged free electrons move across the conductor to the positive terminal of the battery. More free electrons are fed into the conductor from the negative terminal of the battery.

Electromotive force and potential difference

The effect of the battery moving electrons through the conductor is known as **electromotive force**. A mechanical equivalent of the battery pushing the electrons through a conductor is an hydraulic pump pushing liquid through a pipe, for example a central heating pump and system. Electromotive force is measured in volts and is often referred to as e.m.f.

Potential difference can be likened to the difference in pressure in a pipe. For water to flow though a pipe there must be a difference in pressure between the pump outlet and the end of the pipe. In an electric circuit a potential difference must exist between two points in a conductor for the current to flow. Potential difference has the same unit as e.m.f., i.e. volts (V).

Electric current

Electric current is a measure of the rate of movement of electrons along an electric conductor. The greater the number of electrons moving through the conductor, the larger the electric current.

Electric current is measured in amperes (A). A current of 1 ampere is equal to a flow of 1 coulomb of charge per second. Since an electron has a very small charge, the coulomb (C) is used, which is equivalent to the charge on 6.24×10^{18} electrons.

9.4 Electrical circuit diagrams

Electrical circuits such as that shown in Fig. 9.1 are normally drawn in schematic form. That is, all components such as voltmeters, resistors, ammeters and so on, are represented by standard symbols.

Current direction

In Fig. 9.1 the electron flow is shown from the negative to the positive terminal. Current direction was decided upon before electrons were fully understood and is accepted as flowing from positive to negative, which is the opposite direction to the electrons. We call this the conventional current.

Fig. 9.1 Simple electrical circuit using standard symbols.

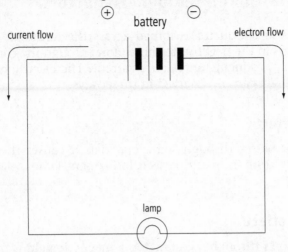

Electrical symbols

Figure 9.2 shows examples of some common symbols representing electrical components which may be used in electrical circuit diagrams.

Fig. 9.2 Symbols used for common electrical components.

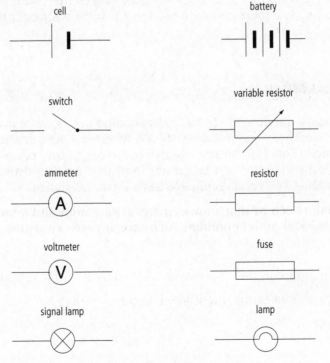

9.5 Flow of charge through conductors

When an electrical conductor is joined across the terminals of a cell, the excess of electrons in the negative terminal flow through the conductor to the positive terminal, producing an electric current. The electric current cannot be seen but its effect can be observed in three ways.

The heating effect

When electricity passes through some materials it causes them to become hot, so that they glow. This can be seen in filament lamps, electric fires and heaters.

The magnetic effect

When current flows through a conductor, a magnetic field is produced. This effect is used in electromagnets, relays, electric motors, bell circuits and electrical measuring equipment.

The chemical effect

An electric current will pass through some liquids known as electrolytes. This produces a chemical change at the electrodes where the charge enters and leaves the liquid. The most common use for this is for electroplating of components.

9.6 Resistance

A material resists the flow of electricity depending on its **resistance**. A good conductor of electricity has a low resistance, whereas a poor conductor has a high resistance. For the same potential difference, the current flowing through a good conductor will be greater than that flowing through a poor conductor because the good conductor has a lower resistance.

If the temperature of a conductor does not change, the current (I) which flows will be proportional to the potential difference applied (V). Thus

$I \propto V$

This is known as Ohm's law and may be proved by a simple experiment.

The resistance, R, relates the quantities V and I such that

$R = \dfrac{V}{I}$ or $V = IR$

The unit of resistance is the ohm, symbol Ω, and it is represented by the letter R in equations.

9.7 Experimental investigation of Ohm's law

Equipment required:

- variable resistor
- ammeter
- voltmeter
- power supply
- a resistor of known value

1 Connect up the circuit as shown in Fig. 9.3.

Fig. 9.3 Circuit used to investigate Ohm's law.

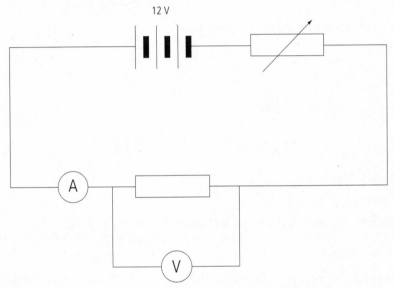

2 Close the switch and adjust the variable resistance to give a current of 0.2 amperes in the circuit and note the voltmeter reading.

3 Repeat step 2 for current values of 0.4, 0.6, 0.8, 1.0, 1.2 and 1.4 amperes and in each case note the voltmeter reading.

4 Plot the graph of V against I and determine the value of the slope.

This value represents the resistance and can be compared with the resistor value.

Figure 9.4 shows the table of results and the plotted graph of V against I.

9.8 Electrical measurement

Figure 9.5 shows a circuit consisting of a power source, a switch, a lamp, a voltmeter and an ammeter.

Fig. 9.4 Results of Ohm's law investigation.

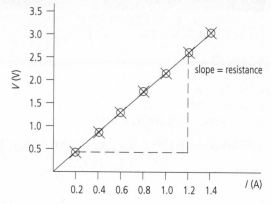

Voltage V (V)	Current I (A)
0.44	0.2
0.88	0.4
1.32	0.6
1.76	0.8
2.2	1.0
2.64	1.2
3.08	1.4

Fig. 9.5 Connecting an ammeter and a voltmeter in a circuit.

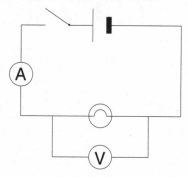

The ammeter measures the electrical current flowing through the conductor. It should be connected in series with the conductor. The ammeter must have a very low resistance as all the current in the circuit passes through it.

The voltmeter measures potential difference and should be connected in parallel with the part of the circuit concerned. To prevent current flowing through the voltmeter from the circuit, it should have a high resistance.

9.9 Series and parallel connections

We have previously seen that components may be connected in series or in parallel. If components are connected in series then they are connected end to end so that the same electric current flows through all the components.

Figure 9.6 shows series connection.

Fig. 9.6 Series connection.

When components are connected in parallel, they are placed parallel to each other as shown in Fig. 9.7. In parallel connections only part of the current goes through each component.

Fig. 9.7 Parallel connection.

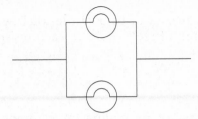

Resistors in series

When resistors are connected in series the total resistance may be found by adding up each resistor value, i.e.

$$R_T = R_1 + R_2 + R_3$$

EXAMPLE 9.1

Three resistors of value 4 Ω, 3 Ω and 6 Ω are connected in series. Determine the total resistance for the circuit.

$$R_T = R_1 + R_2 + R_3$$
$$= 4 + 3 + 6$$
$$= 13$$

The total resistance is 13 Ω.

EXAMPLE 9.2

Determine the total resistance in a circuit made up of two resistors, of value 20 Ω and 12 Ω placed in series.

$$R_T = R_1 + R_2$$
$$= 20 + 12$$
$$= 30$$

The total resistance is 32 Ω.

EXAMPLE 9.3

A circuit has two resistors in series, one 700 Ω and the other 2.2 kΩ. Determine the total resistance.

$$R_T = R_1 + R_2$$
$$= 2200 + 700$$
$$= 2900$$

The total resistance = 2900 Ω or total resistance = 2.9 kΩ.

Resistors in parallel

When resistors are connected in parallel the total resistance is found from the reciprocal of the sum of reciprocals for each resistor value. The formula used is:

$$\frac{1}{R_T} = \frac{1}{R_1} + \frac{1}{R_2} + \frac{1}{R_3}$$

The solution can be reached by finding the common denominator for the reciprocals on the right-hand side and adding the fractions, or by using the reciprocal function on a calculator.

If a circuit has two resistors in parallel, R_1 and R_2, the total resistance can be determined from the product sum, i.e.

$$R_T = \frac{R_1 R_2}{R_1 + R_2}$$

EXAMPLE 9.4

Three resistors of value 5 Ω, 3 Ω and 6 Ω are connected in parallel. Determine the total resistance.

$$\frac{1}{R_T} = \frac{1}{R_1} + \frac{1}{R_2} + \frac{1}{R_3}$$

$$\frac{1}{R_T} = \frac{1}{5} + \frac{1}{3} + \frac{1}{6}$$

The common denominator on the right-hand side is 30, so

$$\frac{1}{R_T} = \frac{6 + 10 + 5}{30}$$

$$\frac{1}{R_T} = \frac{21}{30} = 0.7$$

$$\therefore R_T = \frac{1}{0.7} = 1.43\ \Omega$$

The total resistance = 1.43 Ω.

The solution can also be found by using the reciprocal function on a calculator.

$$\frac{1}{R_T} = \frac{1}{5} + \frac{1}{3} + \frac{1}{6}$$

$$= 0.2 + 0.333 + 0.167$$

$$= 0.7$$

$$\therefore R_T = 1.43\ \Omega.$$

EXAMPLE 9.5

Find the total resistance for two parallel resistors of value 6 Ω and 4 Ω. Remember, if a circuit has two resistors in parallel, R_1 and R_2, the total resistance can be determined from the product sum, i.e.

$$R_T = \frac{R_1 R_2}{R_1 + R_2}$$

$$R_T = \frac{6 \times 4}{6 + 4}$$

$$R_T = \frac{24}{10} = 2.4$$

The total resistance is 2.4 Ω.

Exercise 9.1

1 Determine the total resistance for the circuit shown in Fig. 9.8

Fig. 9.8 Two resistors in parallel.

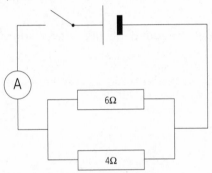

2 A circuit has two resistors each of 12 Ω placed in parallel. Determine the total resistance in the circuit.

3 Two resistors of 6 Ω and 12 Ω are connected in parallel to a 12 V d.c. power source. Determine the total resistance and the current flowing in the circuit.

4 Determine the total resistance for the following circuit (Fig. 9.9).

Fig. 9.9 Three resistors in parallel.

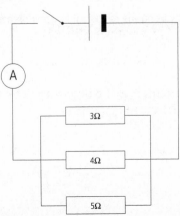

5 Determine the total resistance and current in the following circuit (Fig. 9.10).

Fig. 9.10 Current and resistance calculation.

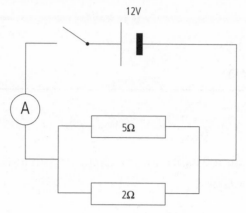

6 Determine the voltage required at the power source for the following circuits (Fig. 9.11).

Fig. 9.11 Voltage calculation for parallel circuits: (a) current of 6 A, (b) current of 2 A.

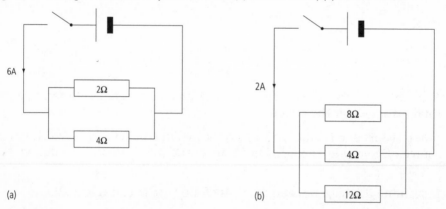

(a)   (b)

9.10 Electrical power and energy

Power

The power in an electric circuit may be determined by multiplying the potential difference (volts) by the current (amps):

power $= V \times I$

Also from Ohm's law $V = IR$

$\therefore$ power $= I \times IR$

$= I^2R$

and since $I = \dfrac{V}{R}$

then power $= V \times \dfrac{V}{R}$

$\qquad = \dfrac{V^2}{R}$

The unit of power is the watt (W).

EXAMPLE 9.6

Determine the power supplied to a 10 A electric kettle connected to a 240 V mains supply.

Power $= V \times I$

$\qquad = 240 \times 10$

Power $= 2400$ W

Energy

Electrical energy is the product of power and time.

Electrical energy $=$ power $\times$ time

The unit of electrical energy is the joule (J).

EXAMPLE 9.7

An electric fire rated at 3 kW runs for 2 hours. Determine the energy supplied.

Energy $=$ power $\times$ time

(note that 2 hours $= 7200$ seconds)

$\therefore$ energy $= 3000 \times 7200$

$\qquad = 21$ MJ

9.11 Fuses

All mains appliances have a fuse in their plugs. The purpose of a fuse is to cut off the current if the appliance passes more current than it is designed for. The fuse is made from a material with a low melting point. If the current is the circuit is too large, the wire will quickly heat up and melt, which breaks the circuit. This is often described as a 'blown fuse'.

It is important that all domestic appliances, i.e. kettles, irons, televisions, washing machines, etc. are fitted with the correct fuse for the particular appliance.

Fuse rating

Fuse rating is found by dividing the power of the appliance by the voltage supply, which gives the current flowing in the appliance. For example, for a 3 kW, 240 V kettle

current = 3000 ÷ 240 = 12.5 A.

The most suitable fuse to use would be a 13 A fuse, which would melt if a current greater than 13 A passed through it.

The most common fuses in the home are 13 A and 5 A.

Chapter review

Electrical conductor   a material that current can flow through, e.g. copper
Electrical insulator   a material that current cannot flow through, e.g. glass
E.m.f.   the effect of an electric cell moving electrons through a conductor
Current   the flow of electrons through a conductor
Potential difference   the measurement of voltage between two points in a circuit
Resistance   the amount by which a material resists the flow of current
Ohm's law   if the temperature of the conductor does not change, the current flowing will be proportional to the potential difference applied
Series connections   components connected in a line, one after the other
Parallel connections   components connected side by side with each other
Power   the amount of energy used per second, units watts (W)
Energy   the product of power and time, units joules (J)
Fuse   a device used to protect electrical equipment from overload and overheating

Exercise 9.2

1 Calculate the power supply to an electric iron which takes a current of 6.6 A when connected to a 240 V supply. Determine the energy supplied in 30 minutes.

2 An electric fire takes a current of 8.3 A when connected to a 240 V supply. Determine the power supplied to the fire.

3 (a) An electric kettle uses 240 V mains supply electricity. If the kettle is rated at 2400 W, determine the value of the current taken by the kettle. If the kettle takes 3.5 minutes to boil, calculate the energy supplied.

(b) State the most suitable size fuse for the kettle.

4 Calculate the power in a two-resistor circuit of 6 Ω and 3 Ω when connected to a 24 V supply when the resistors are connected:

(a) in series

(b) in parallel.

5 Calculate the current in a 60 W table lamp, connected to 240 V mains electricity.

6 A 1.5 kW electric fire is operated from 240 V mains electricity. Calculate the current used by the fire. Determine the energy supplied in 3 hours.

Activity sheet

Series and parallel circuits

Conduct an experiment to investigate series and parallel circuits using the equipment below:

- 12 V power supply
- ammeter
- voltmeter
- connecting wire
- lamp bulbs

Set up the equipment as shown in the three circuits in Fig. 9.12, switch on the power and record the voltage (p.d.) across the lamp bulbs and the current in the circuit.

Fig 9.12   Investigation of three circuits: (a) one bulb, (b) two bulbs in series, (c) two bulbs in parallel.

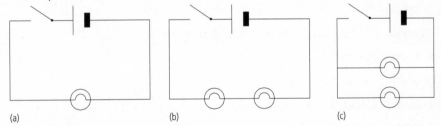

(a)   (b)   (c)

Conclusion

1 Produce a table showing the values obtained.

2 Remove one bulb from circuits (b) and (c) and note the effect. Give the reasons for this.

3 Observe the brightness of the bulbs in circuits (b) and (c), in comparison to that in circuit (a). Give the reason for this.

4 The brightness of the bulbs could be further varied by using a variable resistor in the circuit. Describe two systems where this type of circuit would be useful.

> **Key skills in application of number and information technology can be demonstrated in this activity and recorded in your portfolio of evidence.**

Multiple choice questions

Note that in GNVQ external tests, a minimum of 70% is required to achieve a pass. There are 24 questions in this test, so 17 correct answers are needed for a pass.

1 Which physical quantity is monitored in the cooling system of a vehicle engine?

 (a) time

 (b) current

 (c) temperature

 (d) length

2 The velocity of a train can be calculated by recording

 (a) mass and time

 (b) distance and time

 (c) distance and acceleration

 (d) mass and acceleration

3 A lathe is supplied with mains 240 V a.c. supply. Which quantity will vary during its daily use?

 (a) the supply voltage

 (b) the supply frequency

 (c) the supply current

 (d) the supply fuse rating

4 Which of the following is most important in a forklift truck design?

 (a) lifting capacity

 (b) the frequency of use

 (c) the speed of operation

 (d) the cost

5 A chisel is heat treated in an oven. What quantity remains constant?

   (a) the time

   (b) the length

   (c) the temperature

   (d) the mass

The following laws are commonly used in engineering science:

   (a) Ohm's law

   (b) Hooke's law

   (c) Newton's law

   (d) The principle of moments

Which of the above laws is used for the following applications?

6 To determine the resistance in a circuit.

7 To determine the extension of a spring.

8 To determine the torque applied to a nut.

9 A spring balance is used to measure

   (a) weight

   (b) temperature

   (c) volume

   (d) length

10 A micrometer is used to measure

   (a) resistance

   (b) diameter

   (c) frequency

   (d) current

11 The table shown shows coefficients of linear expansion for four materials

Material	Coefficient of linear expansion (per °C)
Brass	20×10^{-6}
Steel	12×10^{-6}
Copper	14×10^{-6}
Aluminium	24×10^{-6}

Which material will expand the most when heated?

(a) brass

(b) steel

(c) copper

(d) aluminium

12 Pressure in fluids can be determined using

(a) $p = \rho g$

(b) $p = mgh$

(c) $p = ma$

(d) $p = \rho gh$

13 A dial test indicator is used to check

(a) pressure in a system

(b) temperature

(c) dimensional accuracy

(d) the torque on a bolt

14 The centre of gravity of a bus should be

(a) on the roof

(b) close to the ground

(c) at the back

(d) on the upper deck

15 Which would be used to tension a nut to a specified amount?

(a) a spanner

(b) a torque wrench

(c) a socket and wrench

(d) an adjustable spanner

16 A barometer is used to measure

(a) resistance

(b) atmospheric pressure

(c) temperature

(d) weight

17 A turning moment is the product of

(a) force and time

(b) force and distance

(c) distance and time

(d) velocity and time

18 The slope of the graph shown represents

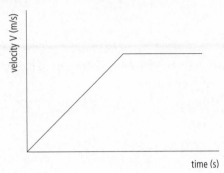

(a) acceleration

(b) velocity

(c) work done

(d) distance travelled

19 Results of a load–extension test for an extension spring are shown in the table.

Load (N)	Extension (mm)
0	0
50	5
100	10
150	15
200	20
250	25
300	30
350	35
400	40

The stiffness of the spring is

(a) 10 N/m

(b) 100 N/m

(c) 1000 N/m

(d) 10 000 N/m

20 The pressure at the bottom of a tank 2 m deep, containing water of density 1000 kg/m³, is

(a) 19.62 N/m²

(b) 19.62 kN/m²

(c) 19620 kN/m²

(d) 196.2 N/m²

21 The total resistance of the three resistors in series is

(a) 571 Ω

(b) 570 Ω

(c) 1570 Ω

(d) 1750 Ω

22 A barometer reading shows 760 mm of mercury. If the density of mercury is 13 600 kg/m³, the pressure in kPa is

(a) 101.9

(b) 101.4

(c) 100

(d) 98

23 The area under a velocity–time graph represents

(a) acceleration

(b) distance travelled

(c) time

(d) work done

24 The current reading on ammeter A would be

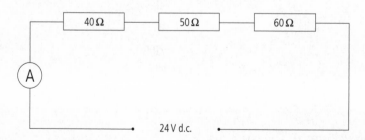

(a) 16 A

(b) 160 A

(c) 0.16 A

(d) 1.6 A

Answers to exercises

Exercise 1.1

1

Quantity	Name of unit	Symbol
Length	metre	m
Mass	kilogram	kg
Time	second	s
Electric current	ampere	A
Thermodynamic temperature	kelvin	K
Force	newton	N
Velocity	metre per second	m/s
Acceleration	metre per second per second	m/s^2
Potential difference	volt	V

2

Multiplication factor	Power	Prefix	Symbol
1000	10^3	kilo	k
1000 000	10^6	mega	M
0.0001	10^{-3}	milli	m
1000 000 000	10^9	giga	G
0.000 001	10^{-6}	micro	μ
0.01	10^{-2}	centi	c

3 **(a)** 0.6 m **(b)** 7.2 kg **(c)** 0.24 km **(d)** 500 mm **(e)** 2.5 tonnes
  (f) 0.65 m **(g)** 1200 g **(h)** 750 mm

Exercise 1.2

1 (a) 7568.9 mm^3 (b) 32.7%

2 Total volume = 31 174 652 mm^3 or 0.031 m^3

Largest volume = diameter 30 mm × length 3 m × quantity 6

= 12 723 450 mm^3

3 3828.5 cm^3 (cc) = 3.8 litres

Exercise 2.1

(a) 21   (b) 7   (c) 5   (d) 8   (e) 9   (f) 20   (g) 15   (h) 75

Exercise 2.2

(a) 20   (b) 35   (c) 45   (d) 30   (e) 18   (f) 45   (g) 12   (h) 28

Exercise 2.3

(a) 23   (b) 17   (c) 39   (d) 46   (e) 15   (f) 10   (g) 11   (h) 21

Exercise 2.4

(a) 15   (b) 26   (c) 60   (d) −5   (e) 5   (f) 36   (g) 36   (h) 1

Exercise 2.5

1 (a) 94   (b) 187   (c) 1190   (d) 1564

2 125 mm

3 160 kg

4 1726 cm

5 1200 kg

6 420 kg; 80 kg

Exercise 2.6

1 Gradient is 2.4 = 2.4 m/s

2 Stiffness = 23 N/mm

3 45.5°C

Exercise 2.7

1 $P = \dfrac{F}{A}$

2 $l = \dfrac{V}{A}$

3 $s = vt$

4 $u = v - at$

5 $t_1 = t_2 - \dfrac{Q}{mc}$

6 $m = \dfrac{y - c}{x}$

7 $a = \dfrac{F}{m}$

8 $t = \dfrac{v - u}{a}$

9 $\text{time} = \dfrac{\text{velocity}}{\text{acceleration}}$

10 $s = \dfrac{v^2 - u^2}{2a}$

Exercise 3.1

(a) 92 kN, 151°   **(b)** 86 kN, 156°   **(c)** 96.5 kN, 21°

(d) 71.3 kN, 97°   **(e)** 96 kN, 113°   **(f)** 58.6 kN, 83°

Exercise 4.1

1 40 kN/m^2 (kPa)

2 188.5 N

3 Extending 530 N, retracting 294.5 N

4 1177.2 N/m^2

5 102.11 kPa

6 100.73 kPa

7 Load = 1500 kN, pressure = 31.8 N/mm^2

8 1.5 mm

9 Load = 30 kN, distance = 0.6 mm

10 2 kN

11 166.67 N

12 33 750 N or 33.75 kN

Exercise 5.1

1 (a) 7.85 N/mm (b) 36 kg

2 0.32 N/mm^2

4 0.9 m

5 (a) 0.5 mm (b) 500

6 80 N

Exercise 6.1

1 (a) 375 N m clockwise (b) 150 N m anticlockwise

(c) 4.5 N m clockwise, 3.75 N m anticlockwise

2 150 N

3 36 N m

4 4 m

5 RL = 13.33 kN, RR = 6.67 kN

Exercise 7.1

1 5 m/s

2 3.2 m/s^2

3 Acceleration = 2 m/s^2, distance travelled = 900 m

4 6 hours 15 minutes

5 Acceleration = 0.64 m/s^2, time = 25.9 seconds

6 Retardation = 1.62 m/s^2, distance = 116.67 m

8 (a) 0.3 m/s^2 (b) 0.2 m/s^2 (c) 60 m (d) 270 m

9 78.48 m

10 (a) 0.33 m/s^2 (b) 0.56 m/s^2 (c) 2040 m

11 981 N

12 61.16 kg

13 300 N m

14 Work done = 3000 J (3 kJ), power = 250 W

15 5886 W

16 9810 W

Exercise 8.1

1 0.003 mm

2 0.734 m or 734 mm

3 3.001 m

7 2.0014 m or 2001.4 mm

8 19.99 m at −20°C, 19.995 m at 0°C, 20.005 m at 40°C

Exercise 9.1

1 2.4 Ω

2 6 Ω

3 Total resistance = 4 Ω, current 3 A

4 1.28 Ω

5 Total resistance = 1.43 Ω, current 8.4 A

6 **(a)** 8 V   **(b)** 4.36 V

Exercise 9.2

1 Power = 1584 W, energy = 2.85 MJ

2 1992 W

3 **(a)** 10 A, 504 kJ   **(b)** 13 A

4 **(a)** 64 W   **(b)** 288 W

5 0.25 A

6 6.25 A, 16.2 MJ

Answers to multiple choice questions

1	c	**2**	b
3	c	**4**	a
5	d	**6**	a
7	b	**8**	d
9	a	**10**	b
11	d	**12**	d
13	c	**14**	b
15	b	**16**	b
17	b	**18**	a
19	d	**20**	b
21	c	**22**	b
23	b	**24**	c

Index